BIRCK BUILDS THE FUTURE

THE FOUNDERS SERIES

The Founders Series publishes books on and about Purdue University, whether the physical campus, the University's impact on the region and world, or the many visionaries who attended or worked at the University.

OTHER TITLES IN THIS SERIES

Purdue Memorial Union: The First 100 Years, 1924–2024
Robert L. Mindrum

Forging the Future: A History of the John Martinson Honors College, 2013–2023
Emily Allen, Jannine Huby, and Pulkit Manchanda (Eds.)

Boiler Up: A University President in the Public Square
Mitchell E. Daniels, Jr.

The Rocket Lab: Maurice Zucrow, Purdue University, and America's Race to Space
Michael G. Smith

Planting the Seeds of Hope: Indiana County Extension Agents During the Great Depression and World War II
Frederick Whitford

Queen of American Agriculture: A Biography of Virginia Claypool Meredith
Frederick Whitford, Andrew G. Martin, and Phyllis Mattheis

WBAA: 100 Years as the Voice of Purdue
Angie Klink

Pledge and Promise: Celebrating the Bond and Heritage of Fraternity, Sorority, and Cooperative Life at Purdue University
Angie Klink

Wings of Their Dreams: Purdue in Flight, Second Edition
John Norberg

Ever True: 150 Years of Giant Leaps at Purdue University
John Norberg

Purdue at 150: A Visual History of Student Life
David M. Hovde, Adriana Harmeyer, Neal Harmeyer, and Sammie L. Morris

Memories of Life on the Farm: Through the Lens of Pioneer Photographer J. C. Allen
Frederick Whitford and Neal Harmeyer

Purdue's Female Founders: The Untold History of Trailblazing Women Faculty
Angie Klink and Jennifer L. Bay (Ed.)

BIRCK BUILDS THE FUTURE

CELEBRATING TWENTY YEARS AT PURDUE UNIVERSITY | 2005–2025

Angie Klink
Zhihong Chen

Purdue University Press
West Lafayette, Indiana

PURDUE UNIVERSITY

Cataloging-in-Publication Data on file at the Library Congress.

978-1-62671-309-3 (hardcover)
978-1-62671-310-9 (epdf)

Cover photo by Charles Jischke.

Design by Natalie Powell.

MISSION

Birck Nanotechnology Center is a world-class research and development facility with the mission to:

EDUCATE

Educating tomorrow's leaders in research and technology

INNOVATE

Driving forward innovations in basic and applied research

COLLABORATE

Forging collaborations with public and private partners

EARLY EVOLUTION OF BIRCK NANOTECHNOLOGY CENTER

1996–2006

The Birck Nanotechnology Center evolved from an interdisciplinary lab into a world-class research facility, playing a pivotal role in nanotechnology research.

1996

JAN FEB MAR APR MAY JUN JUL AUG SEP OCT NOV DEC

- FEB: EE Solid-State Faculty Discuss Renovating Labs in the EE Building
- JUN: Estimate of $4M to Renovate 8,000 SF in EE Building
- JUL: EE Faculty Presentation to Dean Schwartz on Lab Needs
- OCT: EE Faculty Request 10,000 SF Cleanroom, 8,000 SF for Labs, & 14,000 SF for Offices
- OCT/NOV: President Beering Approves Space for Solid-State in New Engineering Building

FUNDING
MEETINGS
CONSTRUCTION
MOVE-IN

CONTENTS

1997

1998

JAN	FEB	MAR	APR	MAY	JUN	JUL	AUG	SEP	OCT	NOV	DEC	JAN	FEB	MAR	APR	MAY	JUN	JUL	AUG	SEP	OCT	NOV	DEC

EE Solid-State Faculty & Staff Gather Information on Research Facilities at Peer Universities, Identify Specific Space Needs at Purdue, and Discuss Location Options with Administration

EE Head Kent Fuchs Asks Dean Schwartz to Include "Purdue Interdisciplinary Nanotechnology Facility" in COE Expansion Plans

$10M Proposal for Congressional Funding for "New Nano Building"

$10M Proposal to State as Part of $40M, 12,000 SF "Indiana Nanoscale Technology Center" (INTEC)

FOREWORD

Of all I learned in my first year at Purdue, the most surprising was the low level of external research funding. In time, it became clear that there was a low level of interdisciplinary, large scale, team-based research. The research support staff in the Office of the Executive Vice President for Research (EVPR) was also relatively small.

Our strategic plan included 300 new faculty financed by a $1,000 tuition increase. Also, we anticipated 600 current faculty would retire or leave Purdue in the six years of our strategic plan for a total of 900 vacancies, which would be filled with faculty who would grow Purdue's research programs.

We also invested $1 million in the Office of the EVPR to increase support for the sponsored research program.

The third aspect of our strategy to grow Purdue's research program was to create a new $100 million platform for large-scale interdisciplinary research, called Discovery Park. We had announced the first building of Discovery Park—a $60 million Birck Nanotechnology Center—thanks to a $30 million gift from Mike, a trustee, and Kay Birck. Also, Dean of Engineering Richard Schwartz and I raised $15 million from alums Don and Carol Scifres for the center. This would be the best nanotechnology research facility at a U.S. university. This complex laboratory took four years to design and build.

In September of 2001, the Lilly Endowment requested a $25 million proposal to "support the strategic vision of Purdue's new president." We asked the endowment to fund $26 million for Discovery Park, including laboratory equipment and seed money for projects. Our concept for Discovery Park was a university-wide interdisciplinary research platform

1999

JAN FEB MAR APR MAY JUN JUL AUG SEP OCT NOV DEC

2000

JAN FEB MAR APR MAY JUN JUL AUG SEP OCT NOV DEC

- MAR 1999: Sagamore Associates Seeks Congressional Funding in DC
- JUL 1999: Vision Expands to Include Physics, Chemistry, Biology, with Engineering
- NOV 1999: Purdue Physical Facilities Estimates $52M for 158,000 SF Building for "Ultra Performance Nanoscale Technology Center" (ULTEC)
- FEB 2000: Dean Schwartz Hires Midwest Cleanroom Assoc. to Perform Budget Study and Create a Marketing Package
- MAY 2000: "ULTEC" Academic Plan Meetings Begin
- AUG 2000: President Martin Jischke Begins Term
- NOV 2000: Letter to VP & Treas. Ken Burns Stating $60M Estimated Cost

consisting of 4 research centers—nanotechnology, bioscience, e-enterprise, and entrepreneurship—which aligned with Indiana's economic development strategy. Discovery Park was funded primarily with private gifts totaling $102.5 million, including $75 million for Birck Nanotechnology Center.

I believe Discovery Park was an unqualified success during my time as President. Half of Purdue's faculty—roughly one thousand—participated in Discovery Park. The interdisciplinary research centers grew from four to twelve. Discovery Park was a powerful asset in recruiting and retaining faculty. Over the first ten years, Discovery Park and its faculty attracted $1 billion of research funding—half the growth of Purdue's research funding. The Birck Nanotechnology Center ranked first nationally in the number of nanotechnology projects funded and second in funding. Discovery Park also contributed to Purdue's growing number of start-ups.

Discovery Park presented challenges, including finding effective leadership for the interdisciplinary research centers and instructors to teach entrepreneurial skills, financial sustainability, and institutional support, as well as renewing its agenda.

But, in spite of these challenges, the first 20 years of Discovery Park and the Birck Nanotechnology Center were major contributors to achieving our goals of increasing research funding and positioning Purdue as a major contributor to Indiana's economic growth.

Happy 20th Birthday, Birck!

Martin C. Jischke

Martin Jischke | President Emeritus, Purdue University

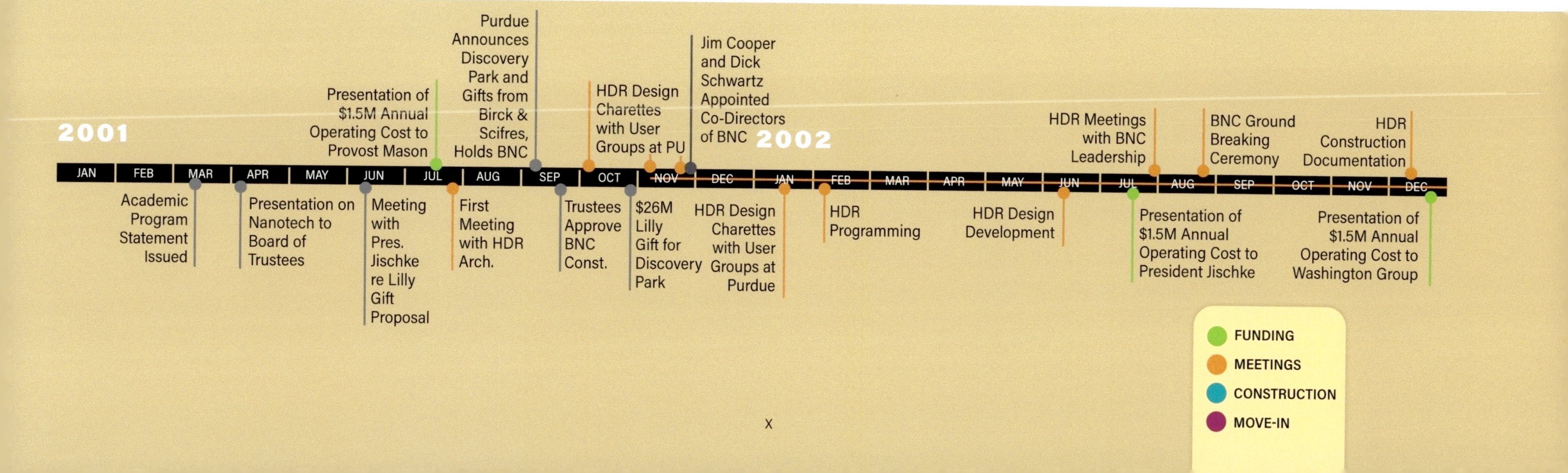

INTRODUCTION

From the Wabash River to the face of the moon, Boilermakers have made a lasting impact across generations. And over the course of the last 20 years, the Birck Nanotechnology Center has helped uniquely position Purdue University for excellence at scale where groundbreaking research, inspirational education, and economic development converge and thrive.

Thanks to the leadership of now-Professor Emeritus James C. Cooper and former Dean of Engineering Richard J. Schwartz, strong support from former Purdue President Martin C. Jischke and funding from Michael and Katherine (Kay) Birck, Don and Carol Scifres, William and Mary Jane Elmore, and Kevin Hall, Birck has paved the way for Purdue to become an international leader in a field that holds revolutionary potential.

Birck's 25,000-square-foot Scifres cleanroom, the finest academic cleanroom in the nation, is complemented by 33,000 square feet of specialized characterization laboratories. More than 300 staff, graduate researchers, and faculty from six academic colleges call Birck their home base, and an additional 300 individuals from our West Lafayette campus, external industry partners, and beyond make use of Birck to achieve giant leaps in nanotechnology. A $50 million investment authorized by Purdue's Board of Trustees, with additional support from the Silicon Crossroads Microelectronics Commons Hub and Lilly Endowment Inc., will enhance and upgrade Birck and its Scifres cleanroom to support the nation's critical need to accelerate the pace of innovation in semiconductor nanotechnology.

In April 2024, we celebrated these renovations with then-National Science Foundation Director Sethuraman Panchanathan and U.S. Sen. Todd Young. The upgrades include 8-inch wafer-scale fabrication and characterization equipment, a training bay for workforce development, enhanced cleanroom infrastructure, new capabilities for quantum research, and advanced packaging equipment to support the SK hynix-Purdue Center for Advanced Packaging, where innovation will rewrite the economics of packaging to make it sustainable in the U.S.

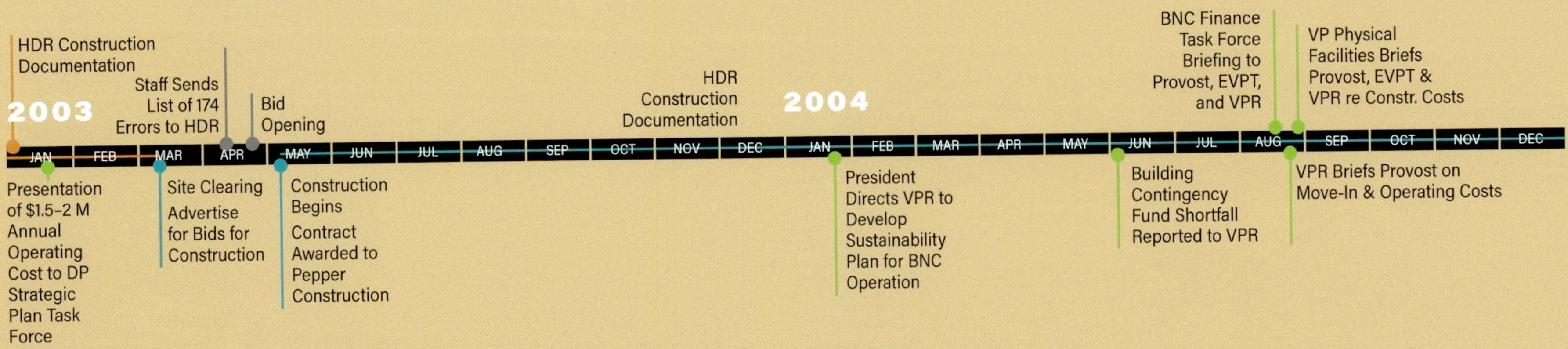

The Birck Center/Scifres Lab will play a major role in the SMART USA Institute, an over $1 billion initiative to develop digital twins that will transform semiconductor design, manufacturing, research and innovation, and talent development. With a $30 million investment by SMART USA and Purdue, Birck's Scifres Lab will become the first digital twin-enabled semiconductor research lab in the nation, one of only two SMART USA Demonstration Facilities, and the only academic facility.

The Birck Center provides students with hands-on learning and fosters collaborations among university researchers, industry partners, and government labs. When university researchers team up with external thought leaders, the potential to produce new ideas, innovations, and solutions multiplies. Students graduate well-equipped to shape America's technology future, ready to translate their classroom and laboratory experiences into real-world solutions for a better tomorrow. Birck's unique digital twin capabilities will support its mission to accelerate the lab-to-fab transition and foster private and public partnerships that grow our nation's semiconductor workforce and R&D ecosystem.

For two decades, Birck has played a critical role in training the engineers and researchers of the future, providing an extraordinary environment where partnerships flourish. The people who have made this center so special over the last 20 years have made a permanent mark on history and will continue to blaze trails for future impact. The recent investments in the Birck Center reflect Purdue University's strong commitment to advancing semiconductor nanotechnology for years to come.

Mung Chiang | President, Purdue University
August 2025

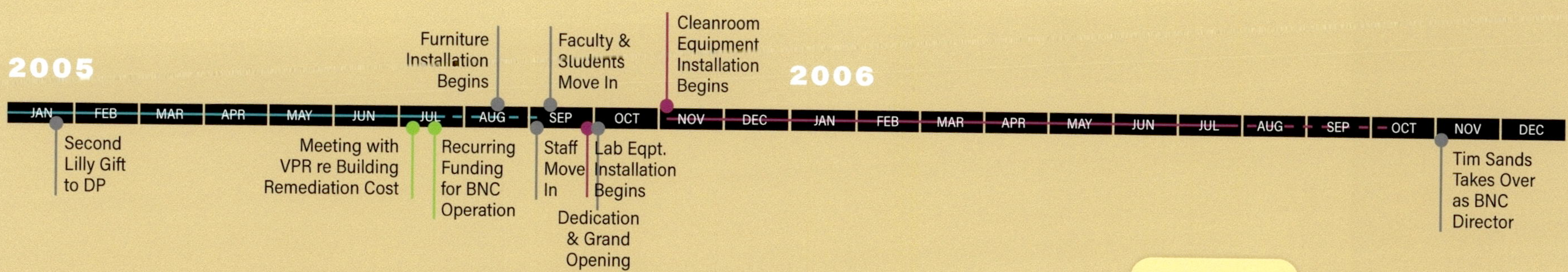

BEGINNINGS LEAD TO DISCOVERY PARK

Initially, the Birck Nanotechnology Center was born out of embarrassment over an antiquated laboratory, the Solid-State Lab, in Purdue University's Electrical Engineering Building. The journey of how it became a groundbreaking interdisciplinary research facility to support the development of integrated circuits key in the establishment of the emerging field of nanotechnology is a story of thinking differently about funding sources and creating the anomaly of exceptional, industry-like laboratories in a university setting.

When the idea for Birck germinated in the late 1990s, nanotechnology was an emerging science in which new materials and tiny structures are built atom-by-atom, or molecule-by-molecule, instead of the more conventional approach of sculpting parts from preexisting materials.

Today at the Birck Nanotechnology Center, electrical, mechanical, materials, chemical, industrial, and aeronautical engineers work alongside physicists, chemists, computer scientists, and biologists to devise structures and processes that utilize the unique features of the nanoscale.

NANOTECHNOLOGY'S IMMENSE INFLUENCE

Nano is a prefix meaning one-billionth—a nanometer is one-billionth of a meter. A human red blood cell is about 7,500 nanometers across, and one nanometer is roughly 10 atoms wide. For example, the letter "**I**" printed here is about 1 million nanometers wide. Just as antibiotics, the silicon transistor, and plastics affected nearly every aspect of society in the twentieth century, nanotechnology is profoundly influencing the twenty-first century.

WAKE-UP CALL

In 1996, then-Electrical and Computer Engineering (ECE) Professor James A. Cooper attempted to recruit two top graduate students to Purdue. However, after touring the Electrical Engineering (EE) Building—now known as the Max W & Maileen Brown Family Hall—and its outdated Solid-State Lab, both students declined his offer. They explained that Purdue's facilities were far too primitive for serious semiconductor research. At the time, the gowning area was simply a public hallway, and once gowned, the researcher

Photo courtesy of Purdue University Archives

The historic Purdue University Electrical Engineering Building, showcasing its classic brick and ivy-covered walls.

The whole concept evolved from our original intention of renovating a handful of about five or six rooms at the EE building. I had a new master's student who had just joined us in 1996 from IBM, and I took him on a lab tour. I was so embarrassed to show him the state of the labs. On February 9, 1996, I sent an interoffice memo to colleagues (Robert) Gunshor, (Jerry) Woodall, (Gerold "Gerry") Neudeck, (Mark) Lundstrom, and (Michael) Melloch (EE faculty) requesting a meeting to make plans to renovate.

James A. Cooper
Retired Founding Co-Director of Birck

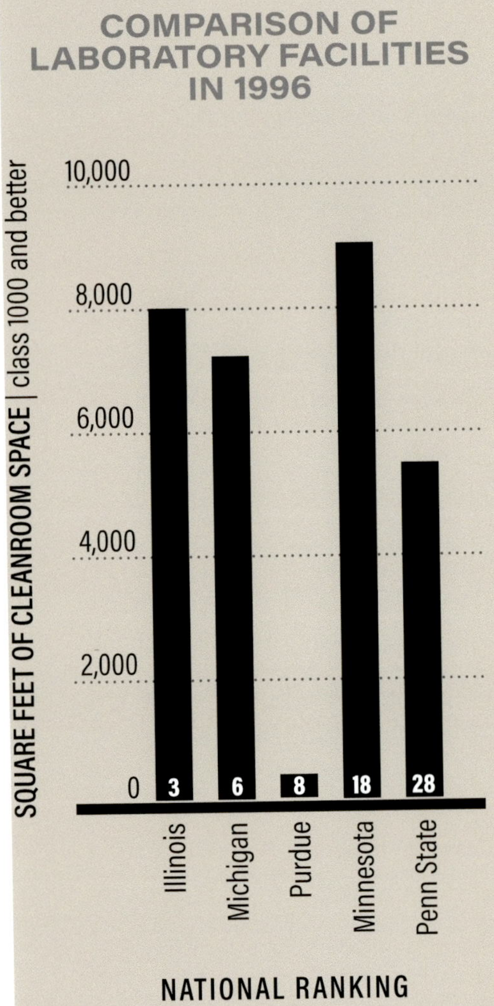

opened the door and merely stepped into the cleanroom. This sparked Cooper to tour and evaluate the facilities at Purdue's sister universities. Upon returning, he escorted another prospective student through the same lab spaces and was again struck by how antiquated the facilities were—unchanged since the 1970s when he had worked there as a graduate student.

Motivated to drive change, Cooper approached then-interim ECE head, Professor David Landgrebe, who toured the labs and immediately shared his concern. Landgrebe encouraged Cooper to reach out to Professor Richard Schwartz—one of the original creators of the lab space—who was then serving as dean of engineering. Schwartz toured the facilities and was equally alarmed. Recognizing the urgency, he brought the matter to then-President Steven Beering, setting the effort in motion. Soon after, Professor Kent Fuchs became the new ECE head and emerged as a strong champion of the vision. He played a critical role in identifying funding sources and learning from the successes and setbacks of peer institutions. Without the aligned and committed support of Landgrebe, Schwartz, and Fuchs, the Birck Nanotechnology Center would not have become a reality.

REALIZATION OF NEED

James Cooper worked with dean of engineering (and former ECE head) Richard Schwartz (who would become his co-director for the Birck Center) in 1996 to form a team to review the existing capabilities and formulate a plan to bring the facilities in line with that of Purdue's sister schools. Other ECE professors and the lab staff who ran the fabrication facilities and cleanroom were enlisted, a rough concept was developed, equipment lists were created, and the infrastructure necessary to support this equipment was determined with support from then-Purdue President, Steven Beering. The team explored state, federal, industrial, and private funding sources.

This work revealed that the existing EE Building could not support the level of necessary infrastructure. Purdue needed a new building. It also became clear that the needs for the facility exceeded the ability of a single department. Planning for a new engineering building began in January of 1998 with a working name for the structure—the Purdue Interdisciplinary Nanofabrication Facility.

1998 MISSION STATEMENT

A document was created dated August 17, 1998, that gave the mission, background, and rationale for what was at the time called the National Interdisciplinary Facility for Nanoscale Technology. Jerry Woodall, Distinguished Professor, School of Electrical and Computer Engineering, and W. Kent Fuchs, head and professor, School of Electrical and Computer Engineering, were listed as contacts on the document. It stated that the mission was: "to establish and construct a nanoscale technology facility that will lead the world in developing ultra small scale and ultra dense information processing technology used in knowledge and distributed intelligence systems. Simply, this facility will enable a new class of user-friendly personal information technology devices."

NAME CHANGES AS MOMENTUM GROWS

Later in 1998 when funds were sought from the State of Indiana, the name changed to the Indiana Nanotechnology Center (INTECH). The hope was that the new facility would act as a magnet to draw business to Indiana. Yet, late in 1999 the name changed again to Ultra Performance Nanotechnology Center (ULTEC). Momentum grew quickly for a new facility and began attracting researchers from across disciplines. Also in 1999, Cooper and Schwartz worked to get biology and chemistry engaged, as this was thought to be an emerging hot area with lab-on-chip technology and the creation of medical devices small enough to be injected into the body.

> *The key to building a preeminent university is the faculty, but without good facilities, you can't attract the faculty.*
>
> **James Cooper** | Jai N. Gupta Professor of ECE, Emeritus

Purdue's unique vision is to put, under one roof in a highly controllable environment, an advanced cleanroom, wet labs, and a technology incubator.

Richard J. Schwartz
Dean Emeritus of Engineering

THE SPIRIT AND PIONEERING INNOVATION OF DISCOVERY PARK

Discovery Park's novel concept of multidisciplinary research and shared space and equipment was groundbreaking. A survey was conducted to determine what other departments would be interested in participating in an advanced laboratory and teaching facility with micro- and nanofabrication capabilities.

Professor of Electrical Engineering (later Electrical and Computer Engineering) David B. Janes, the research program coordinator, organized faculty groups to help understand the needs of the proposed center. Much time was spent on how to manage competing requests for limited resources, how to determine who was "in" and who was "out," how to move researchers out if they became unproductive, and how to attract faculty accustomed to having their own private, locked space.

The effort to expand participation across the university took many forms, including presentations on nanotechnology to faculty in different areas. The response was positive, with mechanical engineering and physics leading the way. Professor of Physics Ron Reifenberger was instrumental in bringing the College of Science on board. Soon several faculty from a myriad of departments participated in the conceptual development of a highly advanced facility.

None of us dreamed that the microelectronics industry would get to where they are now. There were gloom-and-doom predictions about how the massive lithography they were doing would hit a brick wall—they couldn't keep doing it anymore. But we saw that this idea of doing microfabrication, nanofabrication, was pushing things ever smaller at the same time.

David B. Janes | Co-Director of Semiconductor Degrees Program

JANUARY 2000

President Bill Clinton announced the National Nanotechnology Initiative (NNI), which provided a framework for nanotechnology research. This initiative continues to be funded and to date has cumulative funding of $45 billion.

Two overriding ideas emerged: shared equipment and interdisciplinary collaboration. The value of both approaches was obvious to meet the needs of the Grand Challenges as set forth by the National Science Foundation. Without funding, however, the project was not really taken seriously by many faculty members.

At this point, Purdue had a new president, Martin Jischke. Cooper presented the concept to him, and Jischke was enthusiastic about the idea. He said this new building idea dovetailed with a multidisciplinary research complex he envisioned for Purdue. Remarkably, President Jischke would make this facility the centerpiece of the new complex that he would call Discovery Park.

The idea of a new facility grew slowly because it required a funding model beyond what Purdue had acquired in the past. Funding of the needed magnitude of $58.6 million involved contributions from the State of Indiana, the federal government, the university, and private donations.

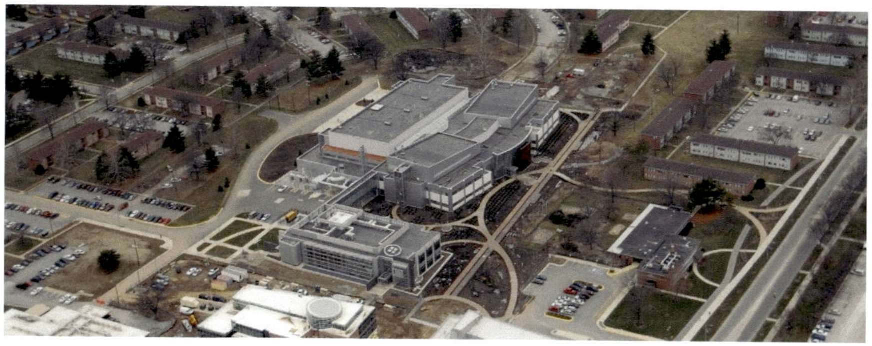

NNI FUNDING BY AGENCY

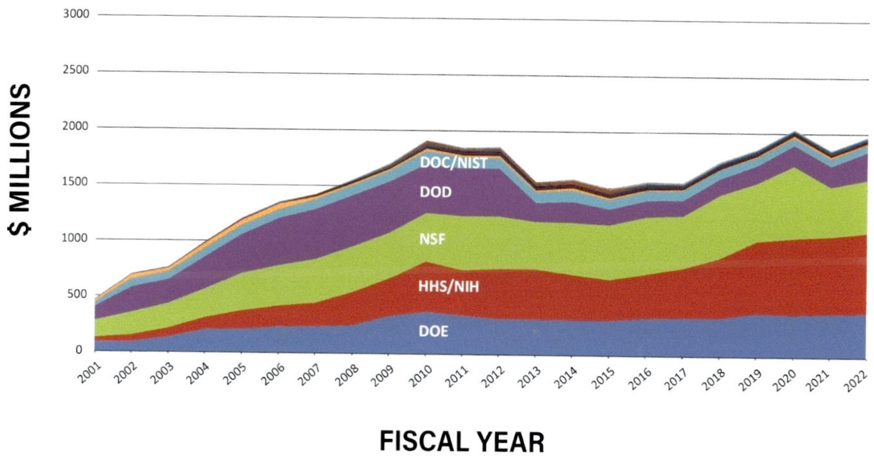

Aerial view of the Birck Nanotechnology Center, showcasing the well-planned layout and design of the Discovery Park District at Purdue University. Photo taken in 2007 by James Cooper and John Weaver.

Katherine and Michael Birck

Photo courtesy of Don Scifres

Carol and Donald Scifres

Mary Jane and William B. Elmore

RECORD-BREAKING DONATIONS AND THE BIRTH OF A LANDMARK CENTER

The creation of the Birck Nanotechnology Center was made possible by extraordinary philanthropic leadership from families whose generosity and vision continue to shape Purdue's legacy in advanced research. In 2001, three transformative gifts—$30 million from Michael and Katherine (Kay) Birck, $10 million from Donald and Carol Scifres, and $2 million from William B. and Mary Jane Elmore—laid the foundation for what would become one of the nation's most advanced academic facilities for nanotechnology research.

Michael Birck, a 1960 graduate in electrical engineering and chairman of Tellabs Inc., and his wife Kay, a nursing leader in Hinsdale, Illinois, had long been champions of Purdue. Michael served as a member of the Board of Trustees and held key leadership roles with the Purdue Alumni Foundation and the Purdue President's Council. Their $30 million gift—the largest private donation in Purdue's history at the time—not only made the nanotechnology center possible but firmly positioned Purdue at the forefront of a scientific frontier that promised to transform fields ranging from computing to medicine.

President Martin Jischke saw how the research enterprise at Purdue was siloed. (Although I don't like to disparage one model over the other.) He realized that what was missing for interdisciplinary research at Purdue was physical space. The vision for Birck grew from that.

Arvind Raman
John A. Edwardson
Dean of the College
of Engineering

> *Virtually all universities with an engineering focus consider nanotechnology to be an essential technology. With this facility, Purdue will be positioned at the forefront.*
>
> **Michael Birck**

In recognition of their remarkable support, the center—originally named ULTEC (Ultra Large-Scale Integration and Technology Center)—was renamed the Birck Nanotechnology Center. The ceremonial groundbreaking in September 2001 launched Discover Purdue Week, which kicked off a yearlong campaign to highlight the university's expanding research capabilities. Construction of the three-story, 187,000-square-foot facility began in summer 2002.

The Bircks were already major Purdue benefactors before their gift to the center. In 1993, they established a $2.3 million endowment in the School of Electrical and Computer Engineering to support undergraduate scholarships, graduate fellowships, and a distinguished academic chair. They also

Katherine and Michael Birck with Neil Armstrong at the dedication ceremony on October 8, 2005.

This new nanotechnology facility will position Indiana to become a player in the "Silicon Valley" of the future ... Purdue already has dozens of top researchers in nanotechnology. They are out to prove that, in this case, if you want to think big, first you need to think small.

Martin C. Jischke
President Emeritus, Purdue University
Taken from a 2001 speech

contributed $3 million toward the Birck Boilermaker Golf Complex and supported the university's aquatic center. Michael Birck received numerous accolades, including the Distinguished Engineering Award in 1991, an honorary doctorate of engineering from Purdue in 1995, and national honors such as the IEEE's Ernst Weber Engineering Leadership Recognition Award.

Alongside the Bircks, Donald and Carol Scifres—both Purdue alumni and Greater Lafayette natives—made a $10 million gift to support nanofabrication capabilities within the new center. Donald Scifres, a pioneering Silicon Valley entrepreneur with more than 130 patents, was co-chairman of JDS Uniphase Corporation and a Fellow of both the IEEE and the Optical Society of America. In recognition of the gift, early reports noted that a wing of the building would be named after his father, Ray Scifres, a member of Purdue's electrical engineering staff for 35 years. However, the final naming decision more fully reflected the central role of the donation: the entire cleanroom facility—one of the largest and most advanced in academia—was named the Scifres Nanofabrication Laboratory.

I received an excellent education at Purdue. The background that I gained in engineering and the sciences was extremely valuable in my professional life. Purdue has meant a great deal to my family also, and we are proud to help Purdue lead the way into this exciting new frontier of science.

Donald Scifres

Like the Bircks, the Scifres family had a long and distinguished history of giving, including a $700,000 donation to preserve parks in West Lafayette. The gift from alumni William B. and Mary Jane Elmore went toward the center's William and Mary Jane Elmore Advanced Wireless Concept Validation Laboratory. The School of Electrical and Computer Engineering (ECE) was named after the family of William B. Elmore in recognition of a $25 million gift in 2021.

Together, the Bircks, Scifeses, and Elmores gave Purdue the resources—and the confidence—to lead in nanotechnology at a time when few universities had made such commitments. Their support enabled a facility with research capabilities unmatched in the academic world, accelerating Purdue's emergence as a national leader in nanotechnology and beyond. Their foresight continues to shape the next generation of technological innovation.

ATTRACTING FACULTY

A call went out to Purdue faculty for groups that could make use of a facility with advanced capabilities. A system was developed whereby different technological areas and departments identified a technology leader to represent them. These leaders formed a team to identify the specific needs and space requirements for each area of the facility. In addition to these leaders, James Cooper reached out to industry partners to garner their input.

AD HOC ADVISOR COMMITTEE

On October 18, 2002, the Ad Hoc Advisory Committee for the Birck Nanotechnology Center met. The committee consisted of Richard Schwartz, dean emeritus of the College of Engineering; George B. Adams III, teaching professor of computer science; H. Hillhouse, professor of chemical engineering; David B. Janes, professor of electrical and computer engineering; Arvind Raman, professor of mechanical engineering (now dean of engineering); Ron Reifenberger, professor of physics; and Tim Sands, professor of materials engineering (now president of Virginia Tech). They participated in a free-flowing discussion on what was meant by "nanotechnology" and gave insights as to the hot topics within their areas of expertise.

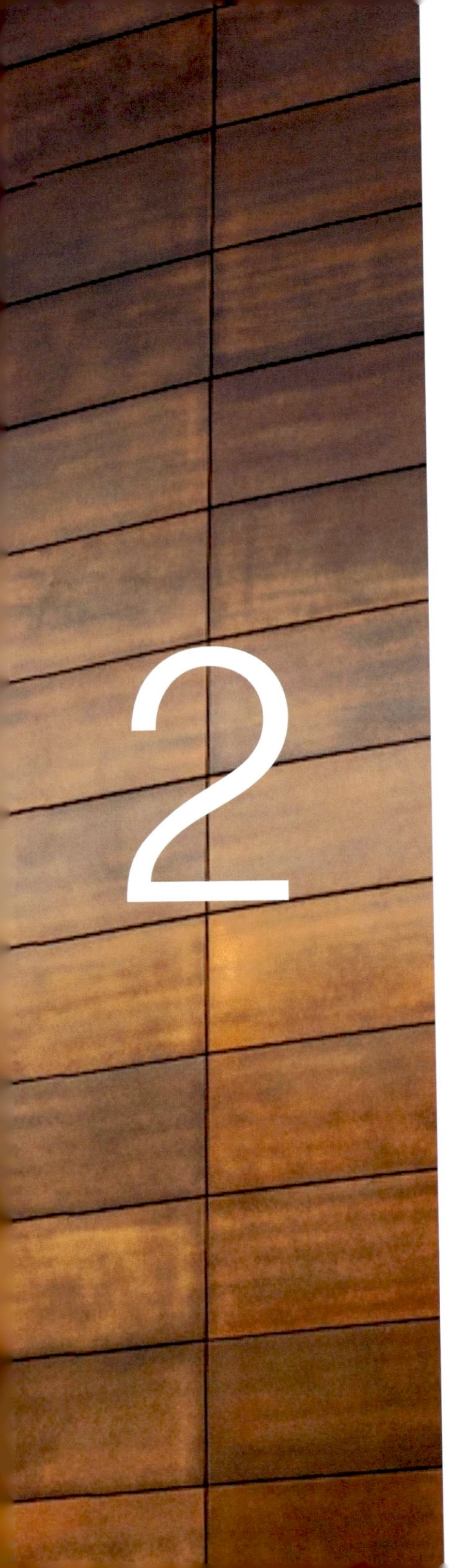

DESIGN AND CONSTRUCTION

ENTER JOHN WEAVER

James Cooper contacted John Weaver, the manager of contamination control for Delco Electronics/General Motors in charge of cleanroom technology, to arrange a tour of Delco's newly upgraded Fab III facility in Kokomo, Indiana. Cooper asked if a one-hour tour would be possible. "If you are building a cleanroom," Weaver replied, "you probably should see all the details that it entails. I would suggest a four-hour tour!"

Cooper agreed, and a tour was arranged for twelve Purdue people—faculty and staff.

After the tour, Richard Schwartz took Weaver aside and said, "We thought we knew what it took to build a cleanroom, but we don't have a clue! Can you help?"

Weaver responded that he was teaching an eight-hour cleanroom design and construction class in California in two months, and he would be happy to do the same class at Purdue. Arrangements were made and about twenty people attended. After the class, Schwartz again took Weaver aside and said, "Now we are really scared! Can you help us?" An arrangement was made for Weaver to spend a "few days per month" at Purdue working on the new facility. He would eventually retire from Delco and work for Purdue full time as the Birck facilities manager.

DESIGN GOALS

The Birck Nanotechnology Center was designed to serve a diverse community of researchers investigating the many aspects of the nanoscale world. Currently there are over 140 faculty members of the center, representing 33 different academic departments across the colleges of agriculture, engineering, pharmacy, science, technology, and veterinary medicine.

The primary design goals for the building were twofold: First, to support a broad range of nanoscale research with the highest-quality laboratory environments possible. Second, to provide a wealth of aesthetically pleasing rooms and spaces to encourage formal and informal spur-of-the-moment, wide-ranging interdisciplinary collaboration among the diverse-background faculty researchers, graduate students, and technical staff who would work in the facility.

DESIGN FIRM SELECTION

The process of designing the Birck Nanotechnology Center began with the hiring of an architectural and engineering (A&E) company with experience in designing a Class 1 (now known as ISO 3) cleanroom. An ISO Class 3 cleanroom represents a highly controlled, ultraclean environment with very strict particle control, limiting particle concentration to a maximum of 8 per cubic meter at 1 micrometer or larger, with no particles larger than 5 micrometers present. It is used in industries such as semiconductor manufacturing, pharmaceuticals, and aerospace, where even the smallest particles can cause issues. The A&E company needed to have experience working with universities and their unique cultural attributes. Advanced laboratory experience was also a key factor, and another important consideration was the methodology for eliciting information from various groups related to their facility needs.

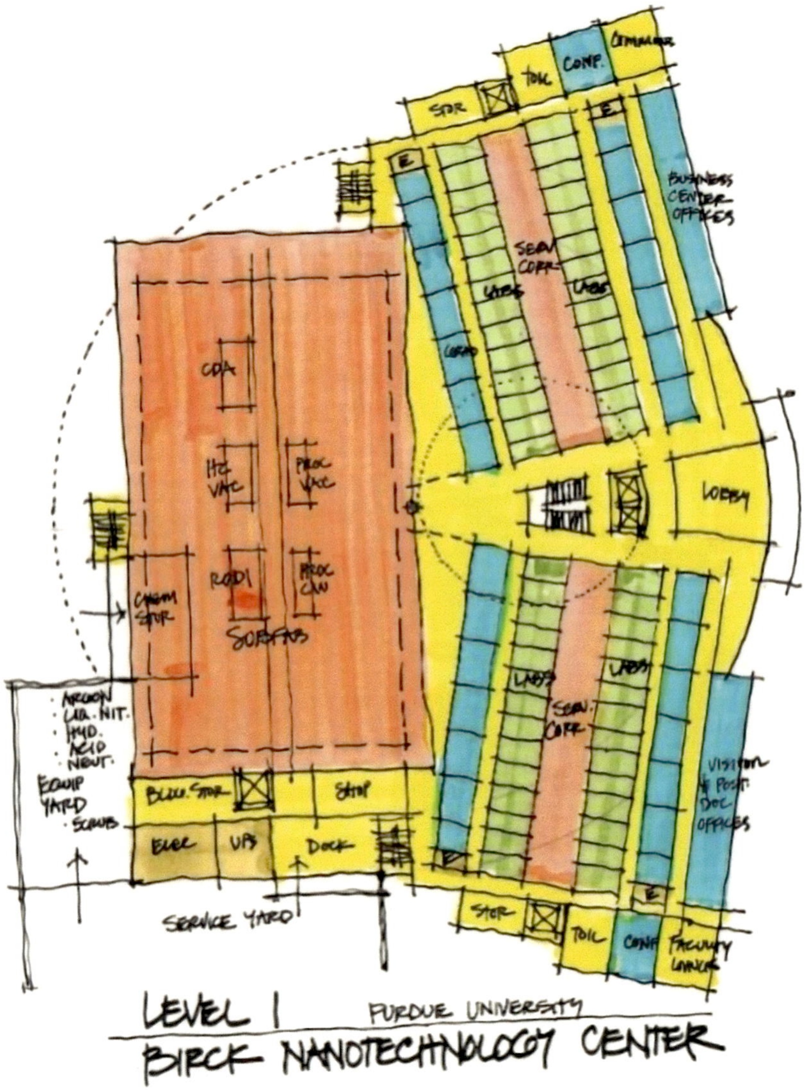

January 2002 schematic design from HDR Architecture.

HDR HIRED

A team was formed by Purdue Physical Facilities' Contract Management Group to pick an A&E firm. This team included James Cooper, Richard Schwartz, and John Weaver, along with people (the users) who would occupy the facility and personnel from Physical Facilities. A list of qualified firms was developed, and four of these firms were invited to Purdue to present their qualifications. HDR Inc., a worldwide architectural engineering firm, was chosen because of their experience designing the new (National Institute of Standards and Technology (NIST) facility in Gaithersburg, Maryland, and their method of obtaining information from the various research groups. It would take two years for the design process to be completed.

DESIGN CHARETTES

For an entire semester, HDR Architecture held a series of multiday meetings, or design charettes, at Purdue with representatives of different research areas including, engineering, physics, chemistry, and biology, and the results were presented back to the groups, sparking further in-depth discussions to iron out specific needs. Discussions included space and equipment requirements, proximity issues, and operating philosophy. George B. Adams III, teaching professor of computer science, became the center's research development manager.

Early rendering of the Birck Nanotechnology Center from HDR Architecture.

OUTSIDE EXPERTISE

The success of the Birck building design arose directly from the willingness of the project management team to seek outside expertise and to collaborate closely with these experts over several years to address the complexities of nanoscale research facility site selection, design, construction, and certification, each of which went far beyond what is familiar for typical building projects.

Because the project was viewed as a regional economic development activity, as well as a signature addition to the university's research infrastructure, it received key donations of employee time from three industry partners: Keithley Instruments Inc. of Cleveland, Ohio, Eli Lilly and Company of Indianapolis, and Delphi Electronics and Safety of Kokomo.

The project management team in the Office of the Architect for Purdue fully integrated the personnel of these industry partners into the large design team comprised of A&E firm personnel and numerous Purdue faculty and physical facilities staff. The input provided by the industry experts was treated with respect and accepted as authoritative. There was an exceptional professionalism to the process.

The project management team rose to the challenges and built the finest nanoscale research facility in academia.

George B. Adams III
Teaching Professor of Computer Science

As the concept of a multidisciplinary facility matured and the various needs were defined, it was decided not to match the sister universities' capabilities, but rather to leapfrog them. The advanced needs of various departments had two effects: it demanded higher capabilities than originally conceptualized and it provided additional funding opportunities.

John Weaver
Cleanroom expert on loan from Delphi Automotive Systems in Kokomo, Indiana

FACILITATING IMMEDIATE FEEDBACK

A key decision was made to include characterization in addition to fabrication, allowing immediate feedback in the research work to be done. The tour of sister facilities revealed that the distance—both physical and cultural—between their fabrication and characterization facilities was both significant and problematic. In one case, the physical separation was more than four miles! The discussions with these universities led to the decision to include characterization in the scope of the facility.

COLLABORATIVE SPACES

The building's many interactive spaces—seating in public areas, small conference rooms, an office design that encourages leaving doors open, open-format student offices immediately adjacent to laboratories, glass vision walls for many conference rooms and labs—foster informal meetings that lead to collaborations. Graduate students report that all the interaction between students and among students from different research projects leads to the rapid diffusion of best practices for experimentation and scientific instrument operation.

ATTRACTING TALENT: HOT AREAS OF THE FUTURE

Facilities like Brick are essential to attract talent. There needs to be alignment between the academic side and the research side of the enterprise for good reason, and interdisciplinary research facilities where there is shared equipment are really important. Birck must continue to reposition itself for the hot areas of the future. And the hot area nationally for the next ten years is going to be chips and quantum.

Arvind Raman | John A. Edwardson Dean of the College of Engineering, 2025

BUILDING PLACEMENT

Once the needs of future users of the facility were determined, a critical decision was made on the placement of the building. A site survey of the proposed location was performed, including testing of the soil system stability, the proximity to vibration sources, and the background levels of electromagnetic interference (EMI).

The soil system, broken rock, was ideal for vibration control, but there was concern that the proximity of the Purdue Airport would cause vibration issues. An analysis was conducted, and it was determined that the landing of airplanes and the acoustic vibration of the aircraft engines were not an issue.

Background EMI levels, however, were problematic. A decision was made to relocate nearby power lines and, when necessary, to bury them underground. These measures brought the EMI levels to acceptable limits.

Concerns regarding the presence of organic airborne contaminants from expended jet fuel from the airport arose, and studies were inconclusive regarding whether these contaminant levels were an issue. To be conservative, carbon filters were installed in the makeup air system to remove any organic matter.

VISION WORKSHOP HELD

HDR completed a preliminary conceptual design for the facility and reviewed it with the technology groups. On November 29 and 30, 2001, HDR held a vsion workshop in which leaders in various fields of nanotechnology were invited to Purdue to review the conceptual design. This event proved valuable as the design concepts were critically examined. As a result of the workshop, a biology component was added.

Site preparation—July (*top*) and August (*bottom*) of 2003.

LEADERSHIP

The team that would occupy the center participated in the design process and led all efforts as the building was constructed. They were led by Dean of Engineering Richard Schwartz and Professor of Electrical and Computer Engineering (ECE) James Cooper. Supporting them were George B. Adams III, David B. Janes, and John Weaver, who was on loan from Delco Electronics.

A construction firm was hired to act as the general contractor, and a separate firm was hired to watch over the construction process and ensure that the contractor performed properly. This oversight proved critical in guaranteeing that the facility was built in accordance with design specifications.

Photos show Birck in various stages of construction from 2003 to 2005.

Entrance, April 2004.

Entrance, September 2005.

Entrance of Birck in 2023.

BINDLEY BIOSCIENCE CENTER CONNECTION

With the expansion of the facility capabilities, a team was created with a technical leader for the nano-bio initiative. Then-Professor of ECE and Bioengineering Rashid Bashir drove the biocleanroom efforts. The biocleanroom and laboratories had two biosafety levels: BSL-1 and BSL-2. It was decided that the use of BSL-3 materials was beyond the scope of this facility, and those efforts would be performed in the nearby Bindley Biosciences Center (BBC). This facility was connected by an overhead walkway to Birck.

The nano-bio laboratories and the biocleanroom were in the east end of Birck, adjacent to the walkway to the BBC. Offices for nano-bio faculty were also located on the second floor of Birck—the level of the walkway connecting the two buildings. This was a deliberate effort to support collaboration between the two centers.

OPEN FOR TOURS

Those involved in the planning realized that the Birck Nanotechnology Center would be a showplace that people would want to see. However, allowing people to tour the cleanroom would compromise cleanliness. The answer was the creation of a viewing corridor and windows surrounding the cleanroom where visitors can watch the operations. Posters next to each window give a capsule summary of the operations taking place in each bay. Safety is an additional advantage to the windows in the cleanroom. If a person in the cleanroom is in distress, they can be seen and attended to.

Left: Skywalk connecting Birck and the Bindley Boiscience Center.

Director Zhihong Chen (*right*) giving a cleanroom tour to His Excellency President Luis Abinader (*left*), President of the Dominican Republic (March 6, 2025).

NATURAL LIGHT

Birck's design was carefully envisioned to maximize the amount of natural light in all offices and labs. High windows provide privacy for the office occupants while allowing natural light to pass through to inside offices. Labs have a wall of windows that can be removed in order to move large equipment in and out.

HIGH-TECH SIDEWALKS

The sidewalks surrounding the exterior of the building project the high-tech character of Birck by incorporating into their designs the Fibonacci ratio, or golden mean (derived from the Fibonacci sequence, a series of numbers where each number is the sum of the two preceding ones). The curved sidewalks emulate the Fibonacci ratio in the descending radii of the curves, while the long straight sidewalk features rectangles in the Fibonacci ratio cast into the concrete.

Second Floor Skylight.

Left: Sidewalk outside of Birck.

Bottom: Looking into a specialized lab for device electrical characterization.

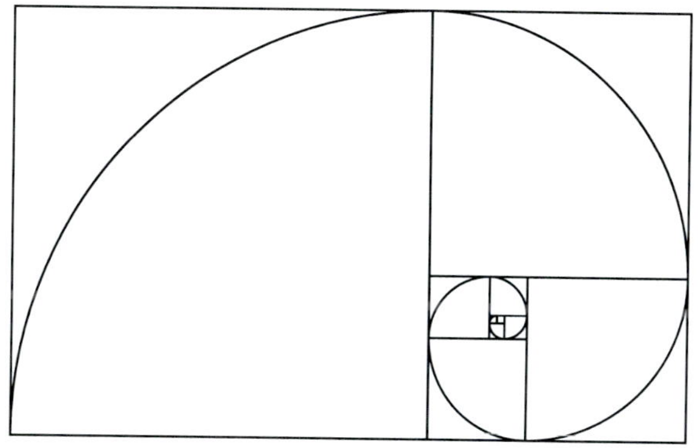

3

SCIFRES CLEANROOM AND HIGH-PERFORMANCE LABORATORIES

The capabilities of the cleanroom, named Scifres Nanofabrication Laboratory, were relatively straightforward; however, deciding what was to be included in the different spaces required great debate. Ultimately the range of capabilities was determined, and an overall concept for the cleanroom was developed. It would be 25,000 square feet of space.

The cleanliness level of the cleanroom was determined to be Class 1 in critical areas, such as lithography, and Class 10 or Class 100 in less-critical areas. With a view to future needs, the cleanroom design allowed for the upgrading of the entire cleanroom to Class 1 if needs and budget drove this level of cleanliness.

The cleanroom would comprise all of the elements necessary for fabrication of the devices and include a teaching laboratory where basic fabrication could be taught and where inexperienced students could use equipment dedicated to teaching. Certain characterization equipment needed for in-process measurements would also be in the cleanroom, but most characterization equipment would be in dedicated laboratories specially designed for this advanced equipment.

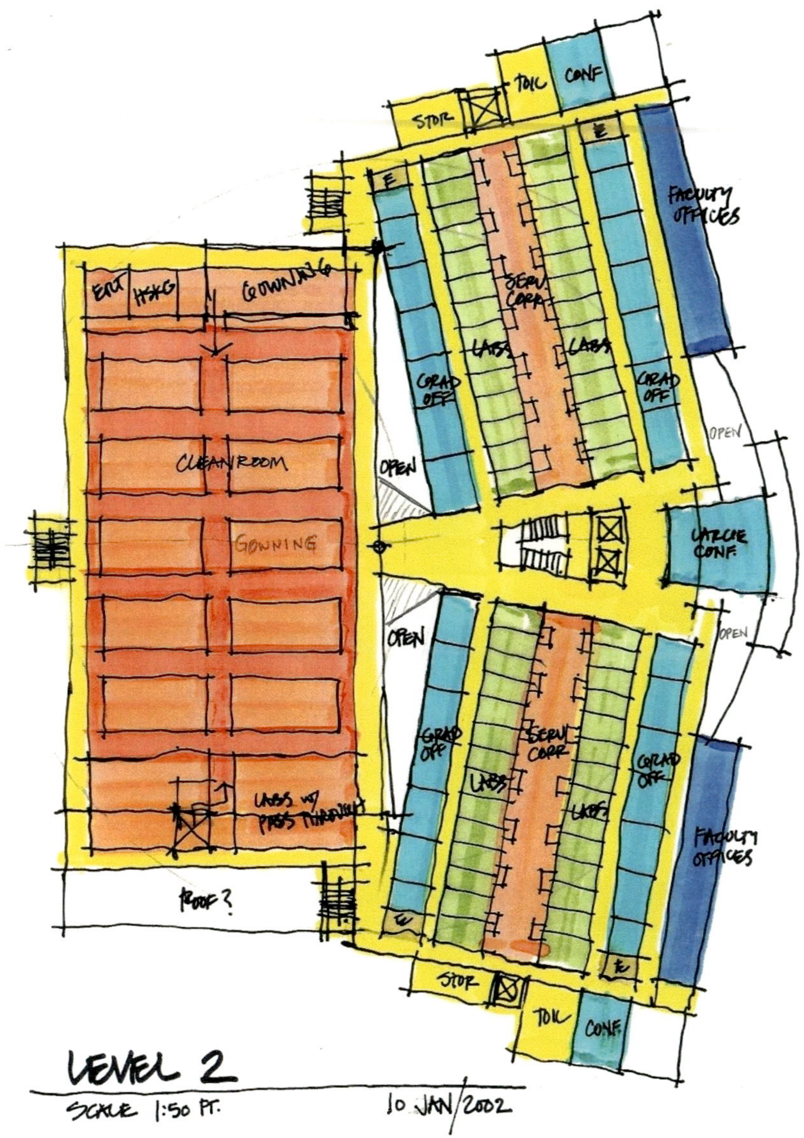

January 2002 schematic design from HDR Architecture.

THREE-LEVEL CLEANROOM

The cleanroom was the most complicated portion of the design. The decision to leapfrog the levels of the cleanroom compared to those of other universities led to the design of a three-level facility. While the three-level cleanroom was standard practice in industry, it was generally considered too expensive for a university facility. On further analysis, however, the need for continuous operation while new equipment was being installed and the speed and reduced cost of installing new equipment justified the higher first cost.

Reduced vibration levels for the second-floor cleanroom are achieved using a waffle slab (from beneath the slab it looks like a Belgian waffle) and close column spacing. An extensive certification validated this approach: The cleanroom exceeded all vibration requirements.

Subfab showing waffle-slab ceiling, which is the floor of the cleanroom.

The lowest level, the subfab, houses support equipment such as vacuum pumps and power supplies that are not cleanable or that generate contamination out of the clean-air stream. Vacuum pumps generate paraffin from heating the pump oil on the outside of the vacuum pump. It was excluded from the airflow path, and therefore the space is not "gowned nor gloved." This allows relatively "dirty" maintenance operations to be isolated from the cleanroom. It also isolates hydrocarbons from vacuum pumps from the cleanroom airstream.

These were important considerations in controlling airborne molecular contaminants. In addition, it had been determined that the hot hydrocarbon byproducts of the vacuum pump oils could condense on wafer surfaces, causing adhesion problems with films such as photoresist. Separating these pumps from the cleanroom air stream mitigated this issue.

Top: Subfab level. *Bottom*: Main cleanroom level.

Top: Main aisle of cleanroom. *Bottom*: Section with perforated floor tiles temporarily removed to reveal the subfloor.

To make the pumps easier to maintain, a set of specialized stands were designed and built in-house. These stands bring the pump up to waist level for easy access, and they contain a drip pan with drain for errant pump oil as well as vibration isolators to prevent the vibration of the pump from being transmitted to the building structure. Bellows units allow the pump to vibrate without transmitting that vibration to the piping leading to the equipment.

The middle level is the cleanroom itself, which comprises three sublevels. The air distribution space is the highest sublevel, ending in terminal ULPA (Ultralow Penetration Air) filters, more efficient than HEPA filters. (HEPA filters are rated 97.6% efficient on 0.3 μm [300 nm] particles while ULPA filters are rated 99.95% efficient on 0.1 μm [100 nm] particles.) These filters are supplied through distribution boxes from the recirculating air handlers.

AIR CHANGES NINE TIMES PER MINUTE

The middle sublevel of the cleanroom is the area where the equipment is located and the researchers work. This area can only be entered through the gowning room, where strict protocol is observed. The air on this level is recirculated at an extremely high rate, with the cleanroom air changing nine times per minute. The clean air provided through the filters flushes the now-contaminated air in the room into the supply side of the recirculated air handlers, where it goes through two levels of filtration before being returned to the ULPA filters and ultimately to the cleanroom.

The lowest sublevel of the cleanroom is the air return level below the perforated floor. This area contains the piping and conduits that supply the cleanroom equipment and the drain piping that carries away waste. The two-foot-square aluminum floor tiles are individually removable for access to this level, and it is tall enough for service people to traverse on hands and knees when necessary.

The highest level of the cleanroom features many large instruments for precision air handling.

CRAWLING TOUR BELOW THE LOWEST SUBLEVEL

When the cleanroom floor was installed, there was some trepidation on the part of the engineering staff as to whether they could traverse this area. To relieve any anxiety, John Weaver led a hands-and-knees tour of this level for all staff members. It was a comedic affair that reassured the staff that the area is not overly cramped.

The highest level of the cleanroom is the air handling level, where both makeup and recirculating air handlers are located. The makeup air handlers are the units that bring in outside air and precondition it for distribution to the cleanroom. This involves setting temperatures and humidity levels, as well as three levels of prefiltration. This air is used to compensate for the air lost to exhaust systems and air leakage from the cleanroom envelope.

Humidity control for this air is achieved through the use of cooling to a set temperature to establish the maximum humidity level and by injecting steam to increase the humidity to a precise level, completely independent of outdoor temperature and humidity. The steam, however, cannot be normal steam that would contain ionic impurities. Instead, ultrapure water heated to steam temperatures is injected into the airflow stream. While expensive, this maintains the purity level of the cleanroom air.

AIR PREFILTERED FIVE TIMES

Air from these makeup air handlers is fed directly into the input side of the recirculating air handlers. These units provide sensible cooling to maintain tight temperature control in the cleanroom. Heating of the air is not needed, becasuse the friction of moving the high volumes of air generates substantial heat. In addition, two more levels of filtration are provided in these units. That means any air entering the cleanroom has been prefiltered five times and is finally filtered through the terminal ULPA filters. This high level of prefiltration extends the life of the terminal filters, which are by far the most expensive filters.

GENERATING AND MOVING AIR, A MAJOR COST

The major cost in any cleanroom is the cost of generating and moving the clean air. The amount of clean air provided, and the cleanliness level of that air, is also the major factor in how clean a cleanroom is. Cleanliness and cost are typically proportional. One means of increasing cleanliness while reducing cost is the use of a bay-chase cleanroom design that maximizes the area of the chase while minimizing the area of the bay. Since the bay is the only area under filter, the amount of air needed to maintain cleanliness is minimized.

The Birck Nanotechnology Center cleanroom utilizes this approach by mounting equipment through the wall that divides the bay and chase. The bay is the only area where research materials—most often silicon wafers—are exposed, and the bulk of the equipment is in the chase. While this is not possible for all equipment, an increasing amount of equipment is designed with this in mind. While this concept had been introduced in industry, it was not commonly used in university cleanrooms. The result is a more efficient operating cost of the cleanroom than that of most university facilities.

Servicing the ULPA filters in the ceiling of the cleanroom working space.

DESIGN GIVES MAXIMUM FLEXIBILITY

The combination of a three-level cleanroom and a bay-chase design gives maximum flexibility to the cleanroom, allowing reconfiguration of the room at minimal cost and allowing the isolation of an area where new equipment can be installed without affecting the remainder of the cleanroom. While this flexibility is helpful in industrial cleanrooms, it is vital to a research cleanroom where equipment change-out is frequent. In the case of Birck, construction of the facility began before the equipment set was finalized, and this arrangement gave a critical timing advantage.

EASE OF MAINTENANCE

The bay-chase design also facilitates maintenance of the equipment. A piece of equipment can be serviced or repaired while the equipment next to it is in operation. The concept of design for maintainability was implemented for each piece of equipment installed. This involved a collaborative effort between the engineer responsible for the equipment and the trades who would maintain the equipment. Both groups collaborated in the design of the installation—what was in the bay, and what was in the chase, how the equipment interacted with the wall system, where the utilities entered the equipment, and so forth.

Crane placing new strobic fans. Strobic discharge fans push a column of laminar air up and into the atmosphere so there is no tall "smokestack" required.

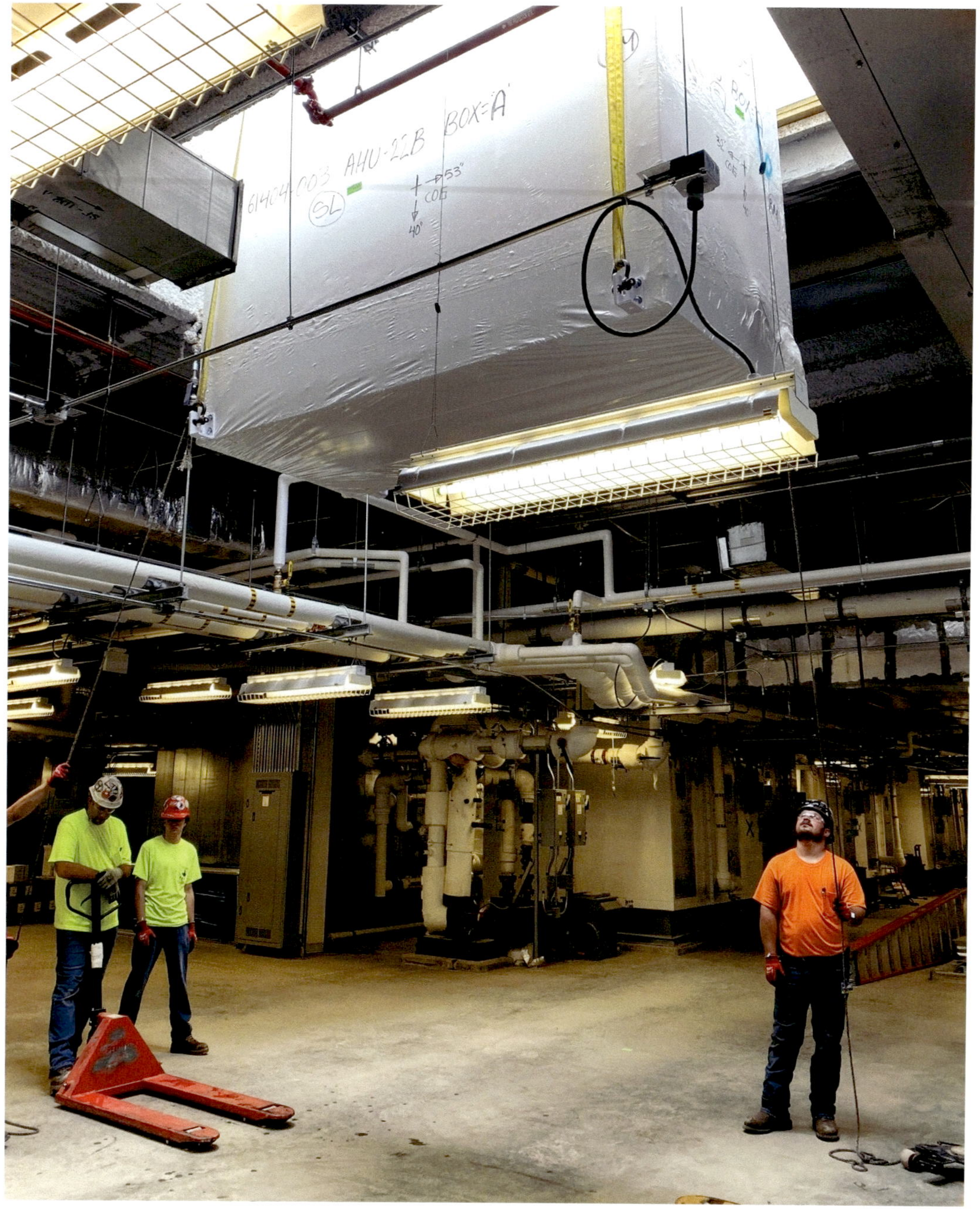

Crane moving a new air handling unit through the roof hatch.

Cleanrooms also require regular manual cleaning.

ELABORATE CONTROLS EMULATED BY OTHER UNIVERSITIES

Many hazardous gases are used in research, and an elaborate set of controls was designed to mitigate the hazards. A toxic gas monitoring system was installed, and all hazardous-gas piping was double-walled with the interstitial between the walls monitored. Distribution boxes are located throughout the subfab, carefully designed and monitored to ensure safety. The safety systems developed for the Birck Nanotechnology Center have been mirrored by many other university facilities around the world.

DESIGNED FOR CLEANABILITY

Design for cleanability was also considered. Each equipment installation ensured that there were no areas around the equipment that could not be cleaned. While the airflow in the cleanroom removes aerosol particles, particles less than 60 micrometers in size, the larger particles settle on surfaces and must be mechanically removed. The elimination of "dead spaces" that would trap particles until something stirred them up, such as a person walking past, was critical. These dead spaces were eliminated in the equipment installation designs.

PREGOWNING

Another factor in the design of the cleanroom was a process supported by the facility design that allowed the movement of people, materials, and equipment into the cleanroom. For people, this process included pregowning and gowning, with the removal of particles from street clothes and cleanroom garments.

A designated area, separated from the rest of the facility with a barrier to passage, is reserved for pregowning. While open to normal room air, this area requires more intense janitorial work. The pregowning area is entered by sitting on a designated bench and, while putting on shoe covers, rotating across the bench such that street shoes touch the floor outside the bench, and shoe covers touch the floor on the pregowning side of the bench. The floor on the pregowning side of the bench is a tacky surface to remove any dust on the shoe covers.

AIR SHOWERS

A person enters the cleanroom, puts on safety glasses, and enters the first air shower—HEPA-filtered air that is blown on the person at a very high velocity. The person then exits into the gowning room, puts on cleanroom garments using a specified procedure that prevents the garment from touching the floor. At this point, the person—now wearing a hood with face mask, jumpsuit, and cleanroom boots—enters the second air shower. This removes any particles that may have gotten on the cleanroom garment during the gowning process. The cleanroom can now be entered.

Pregowning area at the entrance of the Scifres cleanroom.

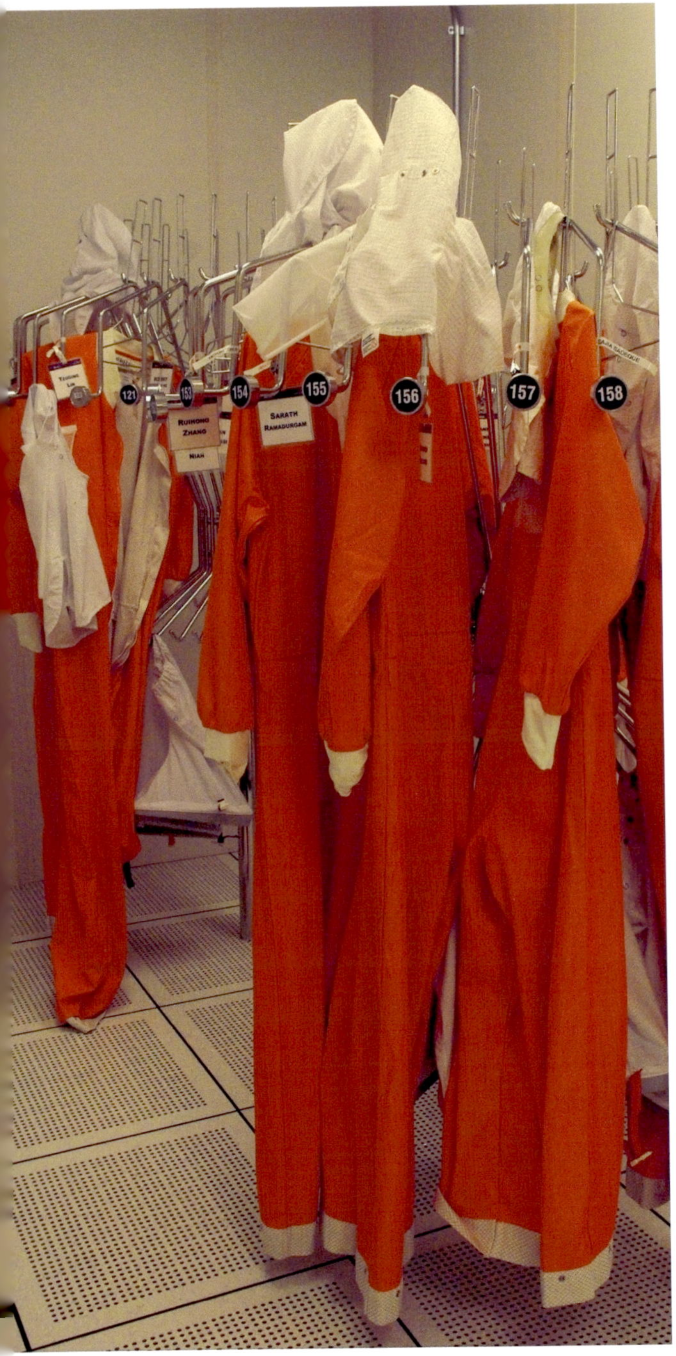

SPECIALTY GARMENTS

Commonly referred to as "bunny suits," the garments worn in the cleanroom are made from GORE-TEX fabric, an expanded PTFE (Teflon) developed by W L. Gore & Associates that creates a strong barrier between street clothing and the cleanroom while providing breathability. Essentially, it is a custom fiber. While common in high-level industrial cleanrooms, these garments are rare in university facilities. W. L. Gore & Associates donated $200,000 worth of cleanroom garments to Purdue for use in the Birck Nanotechnology Center.

MOVING MATERIALS AND EQUIPMENT INTO THE CLEANROOM

The process of moving materials into the cleanroom includes removing any shipping and packing materials from the cleanroom entry point. The material then moves to an area where the outer bag is removed or, if there is no outer bag, it is wiped down. The material then goes into a dumbwaiter that takes it to the second floor in a designated area of the cleanroom. An inner bag is removed in the cleanroom or, if there is no inner bag, the material is cleaned using only cleanroom cleaners and wipers.

Equipment follows a similar process, but on a larger scale. The equipment is unpackaged and brought up to the cleanroom level on a large freight elevator. The initial cleaning takes place in the elevator lobby, then the equipment moves across the hall to the wipe-down area. As soon as it enters this area, the people moving the equipment exit into the hallway and enter the cleanroom through the pregowning and gowning process. They then return to the wipe-down area and fully clean the equipment using cleanroom cleaners and cleanroom wipers. The equipment can then be moved into the cleanroom.

An alternative path was designed for equipment too large or too heavy for the freight elevator. The large equipment is moved from an outdoor crane pad through a demountable wall section in the building exterior into the elevator lobby. From there, the normal path is followed.

GORE-TEX cleanroom garments hanging in the gowning area.

HANDLING EQUIPMENT REPLACEMENT

All equipment will fail over time, and the air handlers are no exception. A hatch is built into the roof of the facility to accommodate the largest air handler section. In the event of a failure, the old air handler can be removed through the hatch, and, more importantly, the new air handler can be lowered into the air handling deck. This hatch would also be employed if additional air handlers were added to the facility.

To facilitate an increase in cleanliness level over time, more air handlers would be required. Cutting large slabs of concrete to allow ductwork to pass through the air handler deck would be a difficult operation. To do so while the cleanroom is in operation would not be possible. To solve this issue, openings were poured into the concrete to allow the installation of enough air handlers to enable the entire cleanroom to be upgraded to Class 1.

"NANO-GRADE WATER"

Rinsing with clean water is most often the last step before a critical operation; therefore the cleanliness and purity of the water is paramount in ensuring a pristine surface. Lab staff member Tim Miller was charged with the creation of a system to generate ultrapure and ultraclean water. By working with ultrapure water suppliers, a multistage water purification system was developed for this highly specialized water. Water with impurity levels measuring in the parts-per-trillion range was generated. This ultrapure water required a new name, and Miller dubbed it "nano-grade water." The name became an industry standard.

Equipment moving through the building en route to its final destination in the cleanroom.

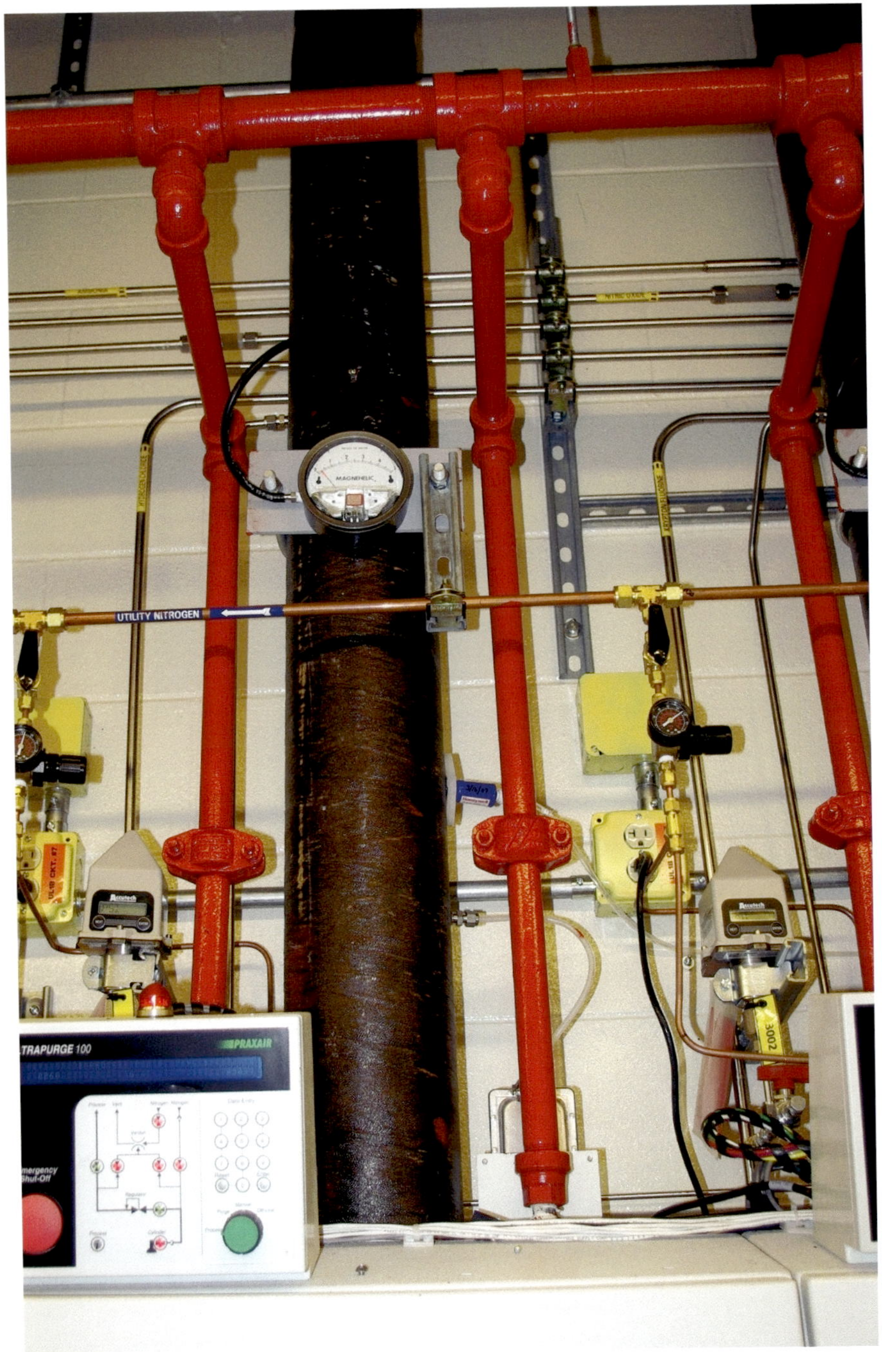

ADHERENCE TO STRICT NFPA STANDARD

The National Fire Protection Association (NFPA) developed a strict fire standard for cleanrooms. While it only applied to manufacturing facilities, it was decided to design Birck in accordance with NFPA 318. This would ensure that the design had the highest levels of safety of any facility of its type. John Weaver was a principal member of that standards committee and oversaw the design goals and ensured compliance to the standard.

COMMISSION PROCESS

As construction was completed in 2004, the commission process began. All building systems were started to be sure they performed correctly. Birck leadership worked with Purdue Physical Facilities to bring the tradespeople who would be servicing the building systems on board during the commissioning process, which was earlier than the normal protocol. This allowed the trades to be part of the debugging of the equipment with the field service representatives from the equipment manufacturers and was extremely effective in providing training and experience. In addition, John Weaver arranged for the tradespeople to spend a day at the similar cleanroom facility at Delco Electronics with the people who supported that facility. Again, the time spent was invaluable.

Gas monitoring system.

CERTIFICATION OF THE CLEANROOM

The final step in the commissioning process was the certification of the cleanroom. A third party was hired for this process, contrary to the normal Purdue protocol. The room was certified at Class 1 (later called ISO 3) for particle contamination, at a temperature control of ±1.0 degree centigrade, and a vibration isolation of NIST A rating. These strict environmental controls enabled highly advanced research and fabrication and set the standard for university cleanrooms around the world.

BENEFICIAL OCCUPANCY

The date that a building is turned over to Purdue by the contractors is called "beneficial occupancy." The week following Beneficial Occupancy, research results were revealed from the first laboratories that were open. That sort of quick research/results in information turnaround time is nearly unheard of. It generally takes many months, and in some cases more than a year, for a facility to have valuable research information to share. Again, Birck set the standard.

The whole concept was to be able to integrate nano-materials with micro/nano-fabricated devices and image structures within these integrated structures, and utilize bottom-up synthesis techniques (directed self-assembly) to realize novel materials and device structures.

David B. Janes
Co-Director of Semiconductor Degrees Program

HIGH-PERFORMANCE LABORATORIES

Additional laboratory needs were determined according to the "thrust areas" of nanotechnology that were supported by the facility. Thrust areas represent the primary research themes emphasized by an institution or lab. The Birck thrust areas, in addition to nanofabrication, were and are nanoelectronics, nanoscale measurements, nanophotonics, and nanoscale imaging. The nano-bio thrust area was added later.

These laboratories are adjacent to the cleanroom to facilitate the rapid characterization of fabricated devices. Laboratory cleanliness is not specified, but temperature and EMI are carefully controlled. All laboratories have a maximum temperature variation of ±1.0 degree centigrade, with tighter control in laboratories with housing equipment that has more stringent requirements. Laboratories that have tighter EMI needs are in the west end of the building, furthest from the electrical distribution systems. The electron-beam lithography systems are in the west end of the cleanroom for the same reason.

In addition to the specifications for all laboratories, certain labs have tighter control over particular variables. For example, the transmission electron microscopy laboratory has tighter EMI and temperature specifications to meet the requirements of this instrument. The lab wing was designed for flexibility with overhead utility carriers, glass walls on labs, and more. Special electrical power distribution/monitoring reduces internally generated EMI.

Scenes from specialized labs in Birck, featuring fume hoods, wet benches, and optics labs with dedicated clean electrical power.

VIBRATION-SENSITIVE LABORATORIES

Vibration-sensitive laboratories are located on the ground floor, with slab-on-grade construction. The maximum vibration is specified at NIST-A levels. This is the same vibration level specified for the cleanroom.

HALL NANOMETROLOGY LABORATORY

Kevin G. Hall gave a generous donation to make possible the Kevin G. Hall Nanometrology Laboratory, a specialized space dedicated to highly sensitive measurements at the nanoscale. It provides a controlled environment with exceptional temperature, vibration, acoustic noise, and EMI. control. A Purdue alumnus, Hall worked at GE Robotics in Silicon Valley. The Hall Lab is built with extreme vibration control, achieved by floating a 30-metric-ton concrete slab on six air pillows, certified at a NIST A1 rating, the quietist vibration rating in existence at that time.

We call them high-accuracy rooms. In these labs we will use scanning probe microscopy and other techniques to examine the structure of materials nearly on an atom-by-atom basis.

George B. Adams III
Teaching Professor Computer Science

Floating slab of the Hall Lab during initial construction.

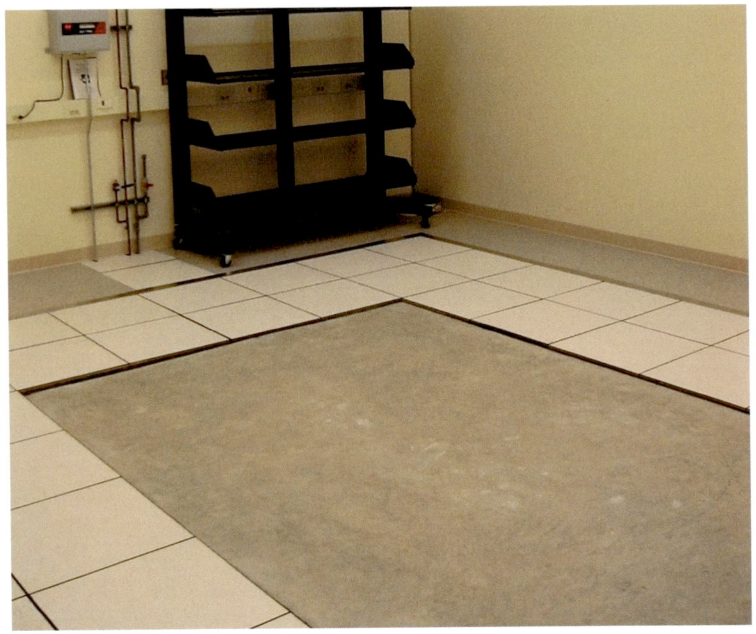

The lab adjacent to the Hall lab has an independent vibration isolation slab and today houses a cutting-edge cryogenic scanning tunneling microscopy (STM) system.

To achieve the high level of EMI control, the walls of the laboratory are ferrous metal with all joints sealed by copper swaged against the walls. The under-slab area and overhead areas are also shielded, completing a full Faraday cage. Electrical inputs to the laboratory go through specialized filter banks to prevent EMI entry. Exhaust ducts exiting the room are copper, also preventing the entry of EMI.

Temperature control is achieved through the implementation of a dedicated makeup-air system and a dedicated recirculating air handling unit. When the room is allowed to stabilize, without people in the room, the temperature is controlled to 0.01 degrees. Once the temperature has stabilized, the experiments are controlled by computer systems located in the support area and connected to the laboratory by a shielded tunnel.

The Hall Nanometrology Lab ready for research.

ENABLING WORLD-CLASS RESEARCH: THE BIRCK FACILITY IN OPERATION

Today, Birck is the research home for over 300 residents, including support staff, graduate researchers, and faculty from six academic colleges. It also hosts over 300 users from Purdue, other universities, and industry. Within the 186,000-square-foot facility are the 25,000 square feet Scifres cleanroom and 33,000 square feet of advanced characterization labs. In addition to the shared use of this cutting-edge research center, the common office spaces for Purdue graduate researchers and visiting scholars encourage interdisciplinary collaborations and interactions.

Right: Scifres Nanofabrication Laboratory.

Resist patterning is a critical step in micro- and nanofabrication, used in combination with deposition and etching processes such as lift-off and etching to define structures. Multiple electron-beam lithography systems, located both inside and outside the cleanroom, enable high-resolution patterning down to 7 nanometers. Optical lithography capabilities within the cleanroom include a maskless direct writer and several mask aligners, supporting patterning from the micrometer scale up to several hundred micrometers. The facility also houses a Nanoscribe photonic 3D printer, which uses two-photon polymerization to print nano-, micro-, or meso-scale 3D objects with high precision.

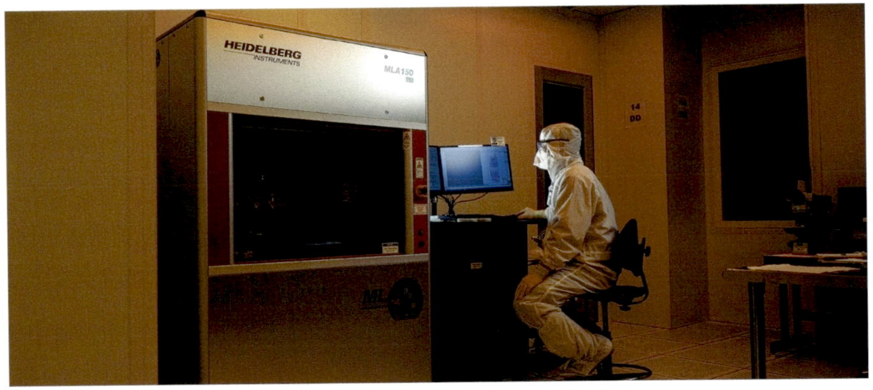

Heidelberg MLA150 Maskless Aligner uses a focused laser for direct-write optical lithography or optical lithography mask creation.

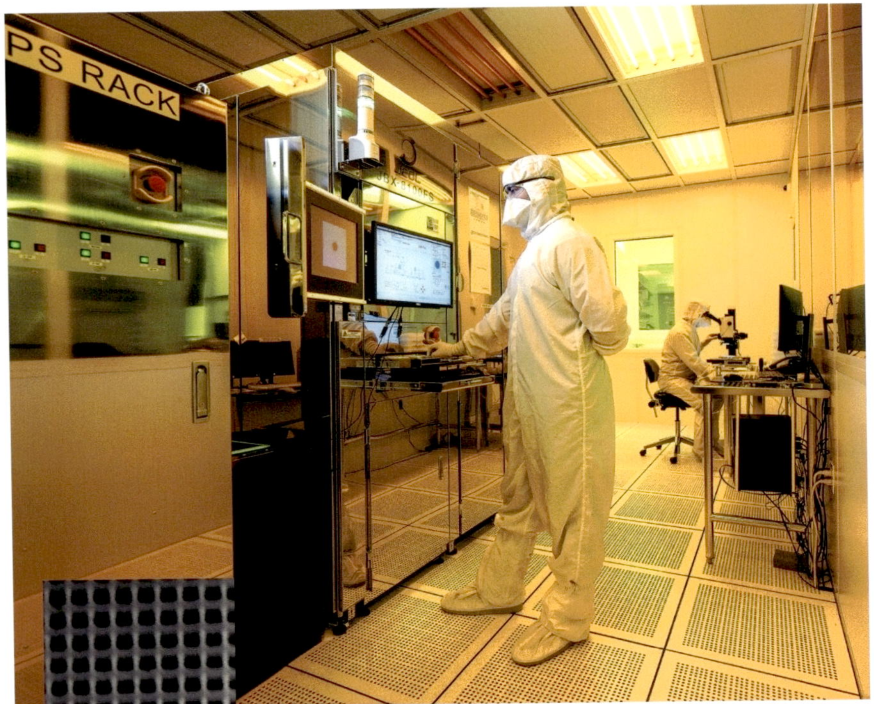

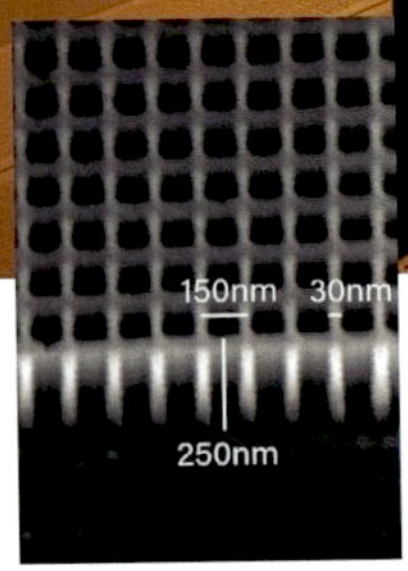

The JEOL JBX-8100FS electron beam lithography system; inset: SEM image showing an example of an etched silicon metasurface that was patterned using the JEOL tool.

The Purdue World's Largest Drum—and world's smallest drum as 3D printed on the Nanoscribe Photonic Professional GT2. The 3D-printed drum is smaller than the diameter of a human hair.

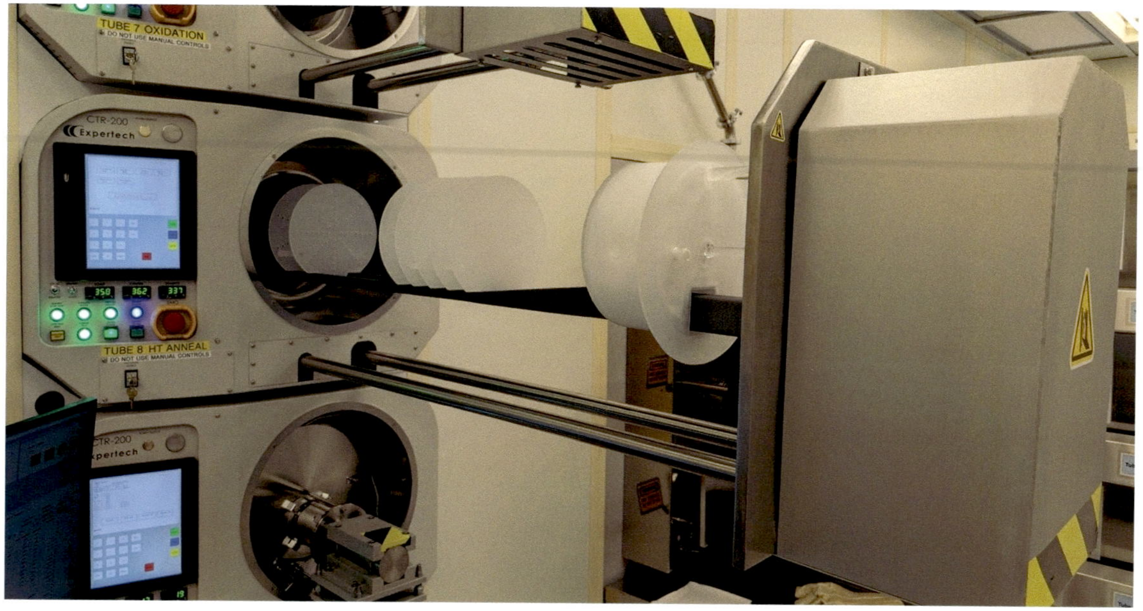

A range of thermal processing equipment is available, including both rapid and traditional thermal annealing. A suite of nine horizontal quartz tube furnaces supports batch processing of wafers up to 8 inches in diameter, enabling various processes such as wet and dry oxidation, chemical vapor deposition (CVD) of nitride, oxide, and polysilicon, as well as forming gas annealing.

One of the nine 8-inch tube furnaces, the tube shown here is dedicated to wet oxidation processes at temperatures up 1,100 degrees centigrade. It can accommodate up to 50 wafers per run, enabling high-throughput thermal processing. Quartz baffles (white) keep the heat and gas flow stabilized in the target zone of the furnace.

SCAN ME

Atomic layer deposition (ALD) glovebox cluster system in Scifres cleanroom.

At Birck, over 30 systems provide a variety of methods to deposit over 60 different metals, dielectrics, and organics. Capabilities include: atomic layer deposition (ALD); chemical vapor deposition (CVD); plasma-enhanced chemical vapor deposition (PECVD); high-density plasma CVD (HDPCVD); e-beam evaporation, DC/RF sputtering, multi-deposition (e-beam evaporation + sputtering); molecular beam epitaxy (MBE); PVD pulsed laser deposition (PLD); and electrodeposition.

Left: Sputtering physical vapor deposition system in the Scifres cleanroom.

Molecular beam epitaxy (MBE) lab.

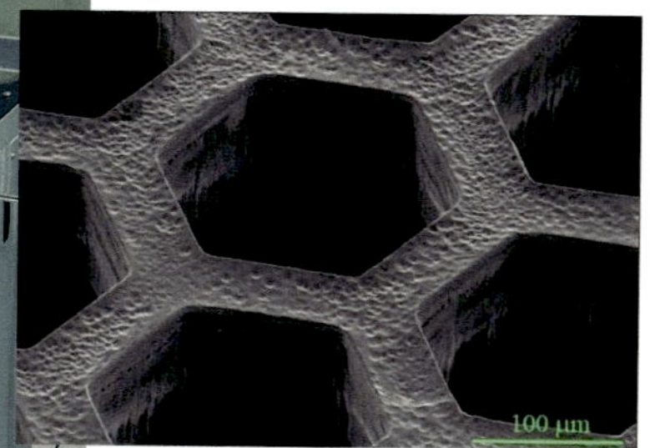

Left: Plasma-Therm advanced etching system in Scifres cleanroom; *right*: SEM imagery showing etch profiles.

In addition to wet etching chemical hoods available inside and outside the Scifres cleanroom, there are many dry etching tools within the cleanroom for reactive ion etching (RIE), deep reactive ion etching (DRIE), asher, and ion milling with a wide variety of chemistries available.

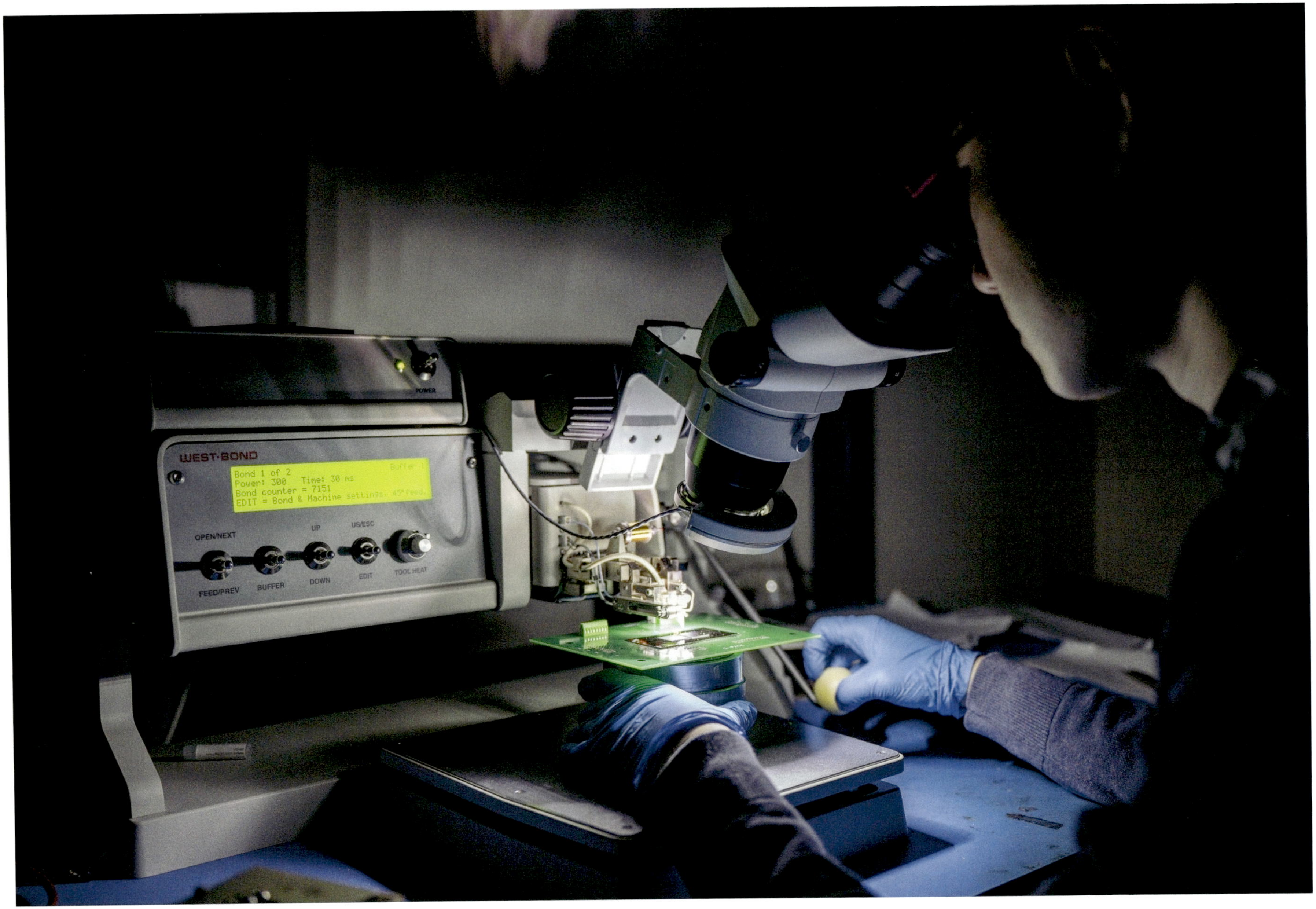

Thermosonic wire bonding.

The Birck Nanotechnology Center supports both wafer processing and packaging capabilities to support device prototyping and evaluation. These include wire bonding for electrical testing, flip chip bonding, wafer thinning, and wafer-level testing. To meet the growing demands of advanced packaging and heterogeneous integration R&D, Birck is actively planning targeted facility expansions to further enhance its advanced packaging capabilities.

In addition to the renowned Scifres nanofabrication cleanroom, Birck is home to a leading-edge facility for materials and device characterization at the nanoscale. All laboratories are designed for low acoustic noise, less than 1 milligauss EMI, and ± 1.0 degree centigrade stability. Additionally, the first-floor laboratories achieve a NIST A vibration rating.

The transmission electron microscope (TEM) laboratory features stringent temperature control, specialized airflow patterns, and acoustic damping materials on walls and floors to ensure a stable imaging environment. The lab houses a Thermis-Z™ microscope, which offers atomic-resolution imaging with resolution reaching below 65 picometers. The system is installed in a high-stability enclosure that provides acoustic, thermal, and electromagnetic isolation, enabling optimal performance. Remote viewing and user control are also supported to enhance accessibility.

Flip chip bonder in advanced packaging lab.

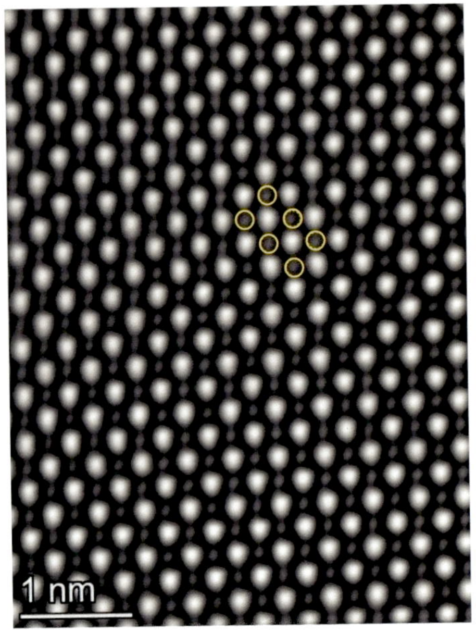

STEM image of $SrTiO_3$ crystal; using integrated differential phase contrast (iDPC) reveals the distribution of the much smaller oxygen atoms in the lattice (circled in yellow).

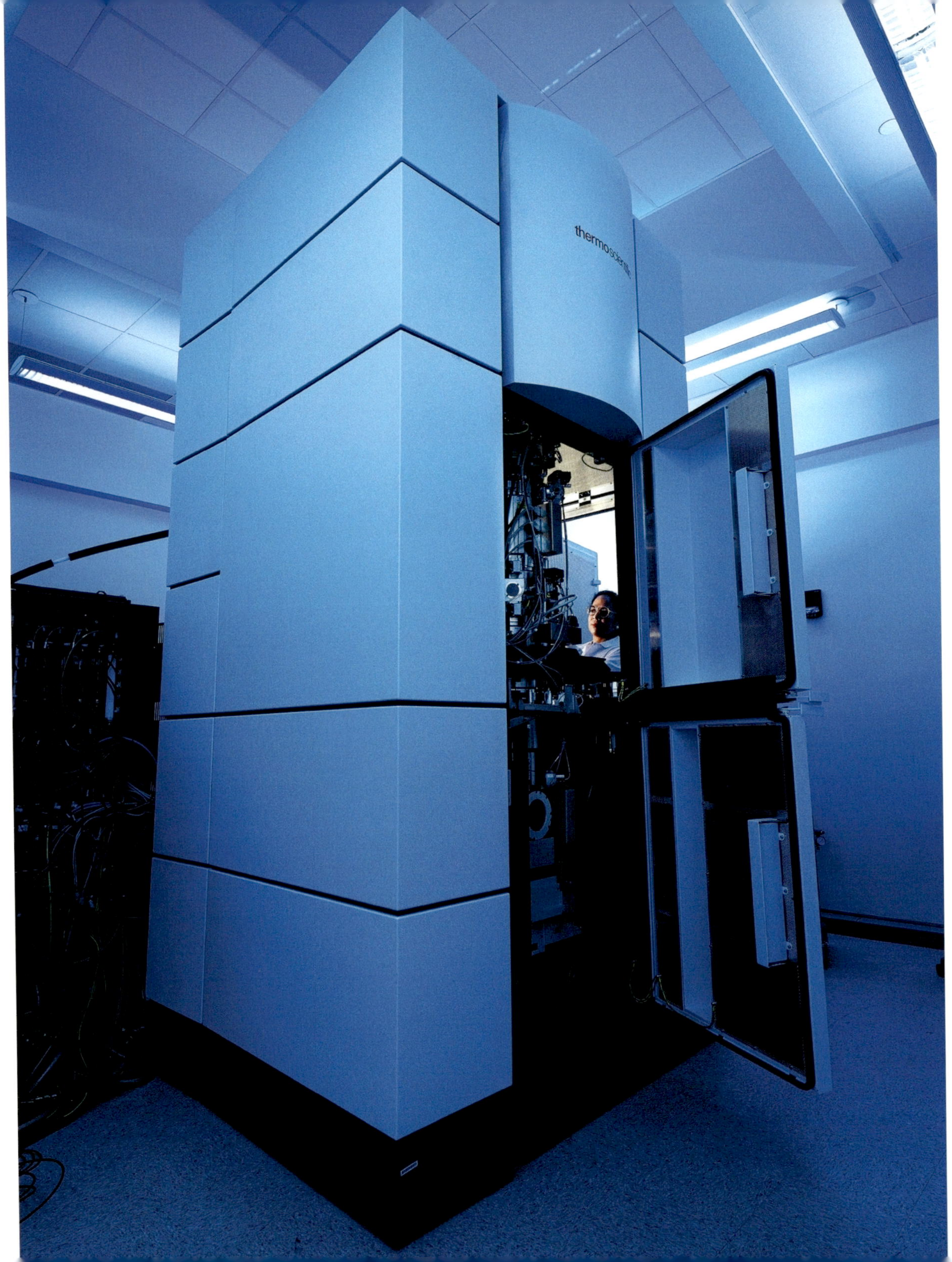

Themis Z aberration-corrected scanning/ transmission electron microscope (STEM).

The Zeiss Orion NanoFab Helium ion Mmcroscope (HIM)/helium-focused ion beam (FIB) uses a focused beam of helium ions to scan across samples, offering exceptional brightness and a narrow energy spread that enables ultrasmall probe sizes. Its source technology, sample interactions, and contrast mechanisms are fundamentally different from those of traditional scanning electron microscopes (SEMs), unlocking a wide range of advanced capabilities—from high-resolution imaging and analysis to precise nanofabrication and characterization. This powerful instrument was made possible through a generous donation from IBM Research.

The Omicron surface analysis cluster is a one-of-a-kind system that combines material growth, processing, and analysis in a single platform under ultrahigh vacuum (UHV) conditions. It allows scientists to study surfaces at the atomic level using powerful techniques such as X-ray photoelectron spectroscopy (XPS), low-energy electron diffraction (LEED), scanning tunneling microscopy (STM), atomic force microscopy (AFM), and high-resolution electron energy loss spectroscopy (HREELS). A central feature of the system is its UHV treatment chamber, where samples can be heated up to 1,000 kelvin or cooled to 120 kelvin using liquid nitrogen. The chamber also supports exposure to various gases, including metal-organic precursors, and is equipped with advanced tools like a mass spectrometer, electron beam evaporator, atomic hydrogen source, metal sputtering source, and multiple thermal evaporators—enabling precise control and modification of material surfaces for advanced research.

Zeiss Orion Nanofab helium ion microscope (HIM)/helium-focused ion beam (FIB).

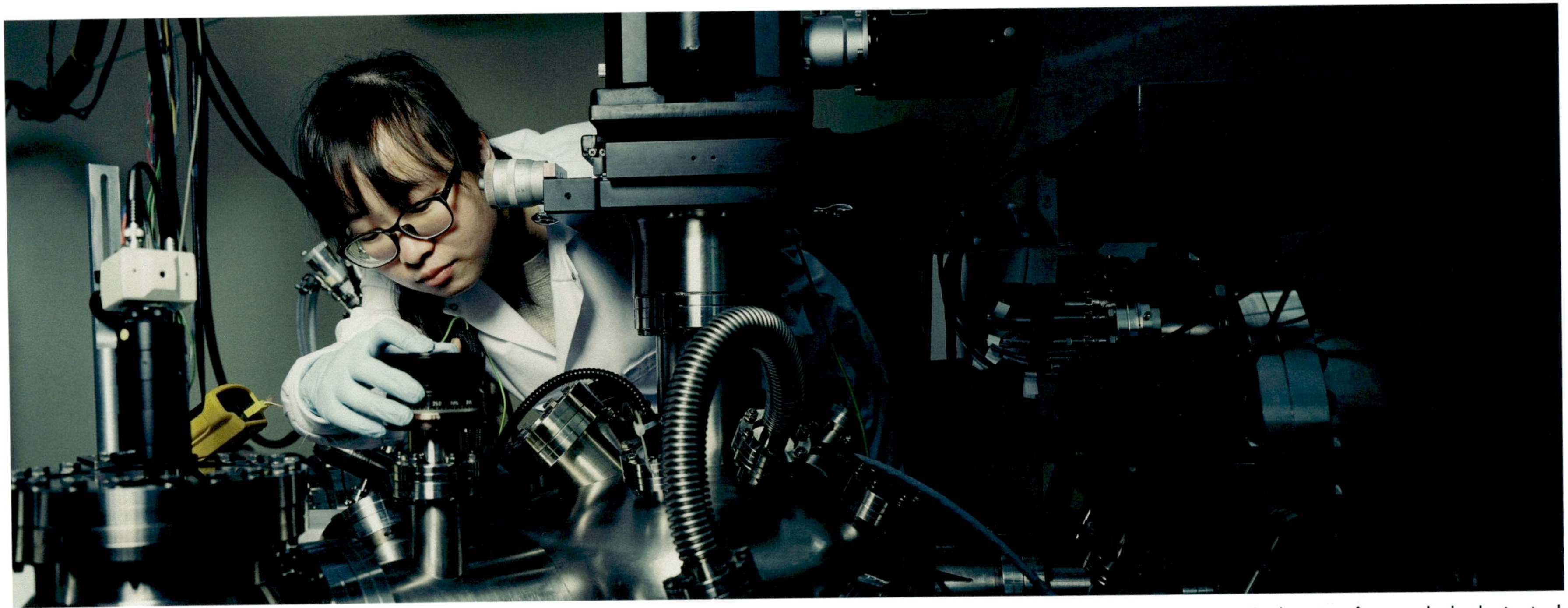

Omicron surface analysis cluster tool.

In addition to its electron microscopy and surface characterization cores, Birck offers a broad set of capabilities for optical, electrical, magnetic, and scanning probe characterization. The optical core includes both far-field and near-field instruments that can reveal material properties with exceptional spatial and spectral resolution. The electrical and gagnetic characterization core features versatile systems that span a wide temperature range and support strong magnetic fields, with probe stations capable of handling large wafers and automated measurements across DC to high-frequency regimes. Complementing these resources, the scanning probe microscopy suite enables nanoscale investigation of surface topography, local conductivity, and mechanical or magnetic properties, equipping researchers with critical tools to advance the development of novel devices and materials.

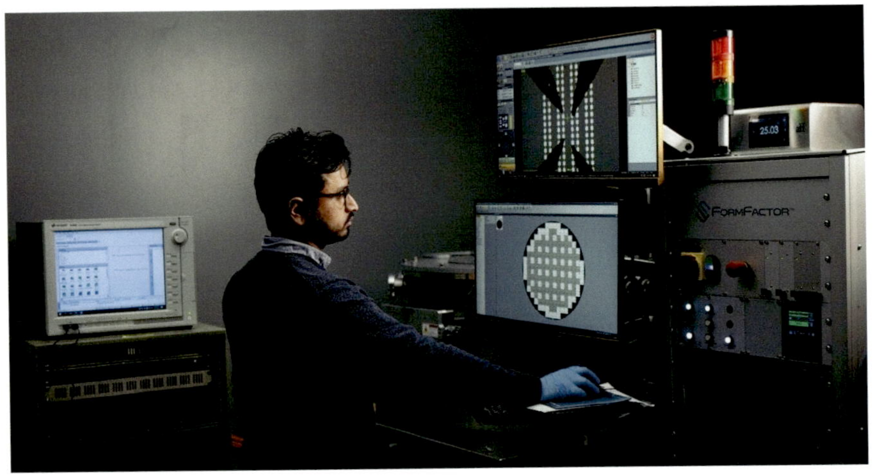

FormFactor PAV200 semiautomated vacuum probe station for 200-millimeter wafer scale measurements.

Right: Inside a cryogenic DC/RF probe station.

132157
P-16-3100-A

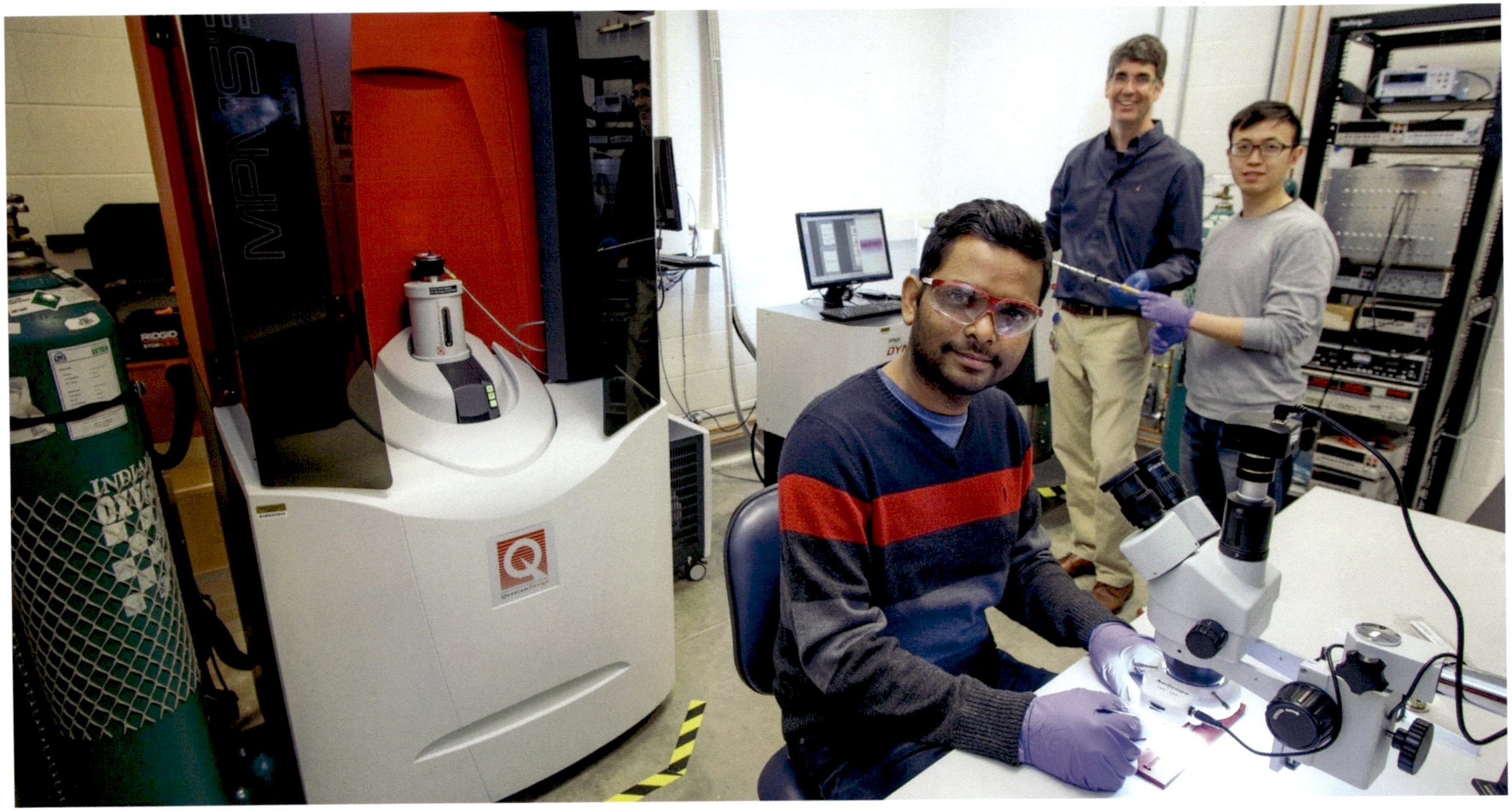

Top: Spintronics lab for cryogenic magnetic, electrical, and thermal measurements.

Right: Nanophotonics lab.

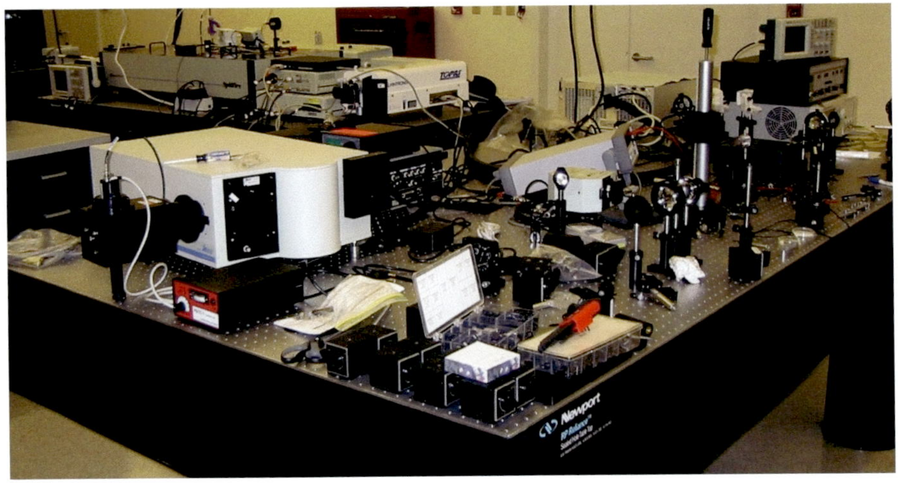

Neaspec IR-NeaSCOPE near-field scanning optical microscope (NSOM) system.

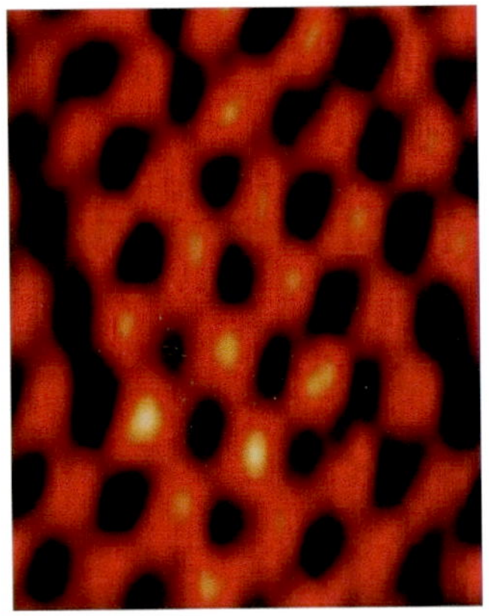

Examples of scanning probe microscopy in the days before Birck. *Above left*: An atomically resolved STM image of the atomically flat surface of cleaved MoS_2 (circa 1996). *Above right*: An AFM image of a long straight multi-walled carbon nanotube (MWCNT) contacted by three separately addressable gold pads (circa 1998). The underlying (dark) substrate is a silicon (Si) wafer. The separation distance between the contact pads is 4 micometers.

This research example is from work done within the new collaborative environment at Birck. A subsurface image of a 1-dimensional polymer composite using an advanced AFM measurement mode: resonance-enhanced 2nd-harmonic kelvin probe force microscopy (KPFM). The sample under study is a composite thin film containing 3 wt% (weight percent) single-walled carbon nanotubes (SWCNTs) embedded in a polyimide (PI) film to form a composite material with 1-dimensional buried subsurface inclusions. In image (c) *below*, the surface topography of the polyimide film provides no evidence of buried SWCNTs. In image (d) *below*, a subsurface KPFM map of the same region reveals the presence of embedded SWCNTs.

(c)

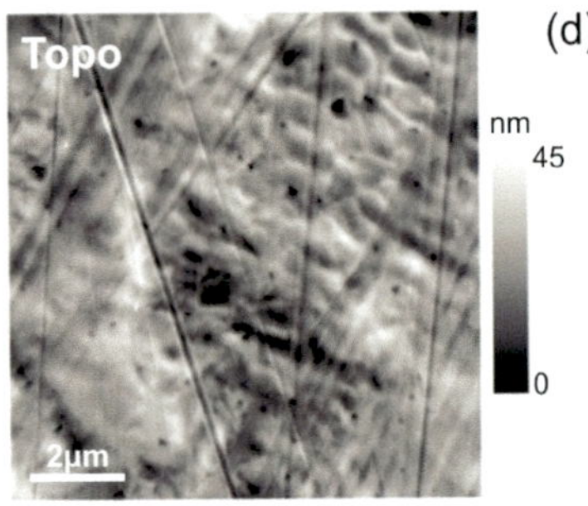

(d)

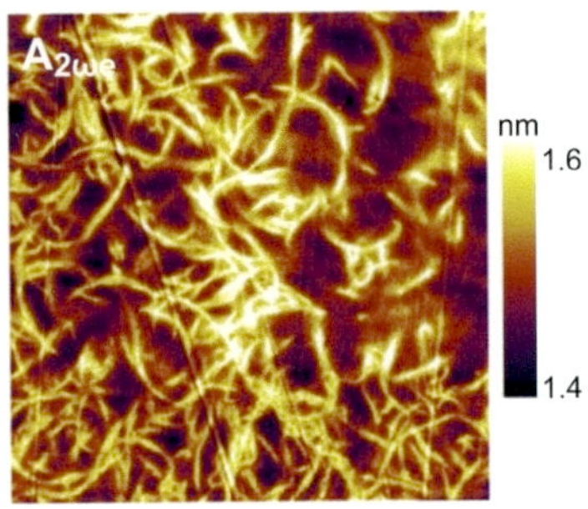

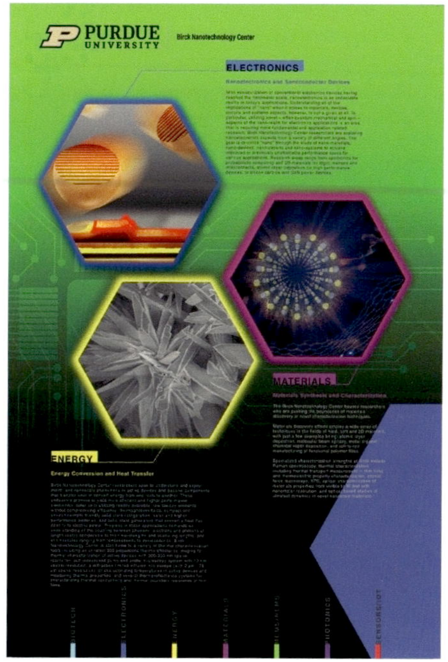

While it is impossible to capture the breadth of research enabled by the Birck Center, the 3 posters above were designed and printed in collaboration with Purdue's Department of Art and Design in 2020 to highlight some of the leading areas of research in Birck at that time. The seven areas that emerged were

MEMS/NEMS | micro- and nano-electromechanical systems, including (radiofrequency) RF devices for inertial sensing, biomedical applications, and wireless interfaces

Photonics | nanophotonic engineering, quantum nanophotonics, and metamaterials

Biotech | bionanotechnology such as bioMEMS and nanomedicine with applications in food security, energy, and environmental solvency

Sensors/IoT | (Internet of Things) including smart manufacturing and printed electronics

Electronics | novel materials, devices, circuits, and computing architectures exploiting the nanoscale and quantum effects;

Materials | synthesis of novel materials as well as development of new nanomaterials characterization techniques

Energy | heat transfer and energy conversion at the nanoscale

INTERDISCIPLINARY RESEARCH BEGINS

DEDICATION

The Birck Nanotechnology Center was dedicated on October 8, 2005. The $58.3 million center involved about 260 faculty and staff members and graduate students from twenty-five schools and departments across the university. Many facility tours were given by students and staff.

Birck Nanotechnology Center dedication 10/8/2005.

I have so many great memoires of Birck opening. I was a tour guide on the day it opened. The dedication was a breakfast event under a tent outside Birck. Orange juice was served in beakers. We all got a real laugh at that. Because the last thing you would want to do is make people feel comfortable drinking out of a beaker. At that time, I began investing my time, which took about six months, into the challenges of transitioning disparate labs across campus into Birck. And honestly, it's one of the ways in which I was able to develop good relationships with the staff.

Aaron Franklin, PhD

Earned a PhD from Purdue University in 2008; currently the Addy Professor of Electrical and Computer Engineering and Associate Dean for Faculty Affairs at Duke University

Breakfast during the Birck dedication 10/8/2005.

TRANSITIONS BETWEEN 2002 AND 2007

BY DAVID B. JANES | Co-Director of Semiconductor Degrees Program

The window between 2002 and 2007 was an eventful time for both the building and the nanotechnology community at Purdue. From 2002 to 2006, Purdue was in the midst of a hiring period. The College of Engineering (CoE) and College of Science (CoS) both had "cluster hires", with nanotechnology representing a key focus. Faculty hired in this period included several who went on to leadership positions within centers or administration, including Arvind Raman (currently dean of engineering), Tim Sands (subsequently Birck director, Purdue provost, and currently president of Virginia Tech), Fabio Riberio (director of CSTAR), Tim Fisher (currently at UCLA), and Eric Stach (currently at the University of Pennsylvania).

This hiring push was somewhat the fulfillment of the planning stages of Birck, with new research areas and new equipment capabilities occupying spaces in the building. In many cases, equipment purchased by these new faculty was funded in part by funds from the Lilly Endowment or institutional matching funds for the Institute for Nanoelectronics and Computing (INAC). Since the Birck building was not slated to be completed until late 2005, new equipment was generally installed in other buildings on campus, with the intent of relocating to Birck when the facility was ready.

The Birck Nanotechnology Center at Purdue's Discovery Park is the epitome of a university-based research facility focusing on the growing discipline of nanotechnology. It is a facility that will become the model for how universities answer critical questions in the exciting and dynamic arena of nanotechnology.

Mihail (Mike) Roco
Senior Advisor for Nanotechnology, National Science Foundation

MOVING IN

Birck main entrance.

The building was under construction from 2003 to 2005, with Pepper Construction as the main contractor. The main construction was completed by the Birck grand opening, in Fall of 2005. At the time of the grand opening, the building was largely empty, although several faculty had started using their offices, including Professor Tim Fisher of mechanical engineering and Richard Schwartz.

Although equipment could not be moved into Birck until the contractor turned over the building, planning for the move had been underway for many months. By this point, a portion of the Solid-State Lab (SSLAB) staff were now Birck engineering staff (including Tim Miller, Dave Lubelski, and John Coy). Mike Longerbone from Ultimate Vacuum Services was hired to assist with the overall scheduling and sequencing.

In order to minimize disruption of research projects, which would have delayed the progress of graduate students and postdoctoral scholars, the staff developed a staging plan in which equipment of a given type (e.g., evaporators, lithography systems, etchers) was moved in sequence, ensuring that researchers had access to a system either within SSLAB or Birck. The process of relocating existing equipment was dovetailed with installation of new equipment, including updates/upgrades to a number of capabilities compared to the existing equipment in SSLAB. While students needed to travel back and forth across campus to perform a complete process, this sequencing was very effective in enabling research with minimum disruption. The overall move-in process lasted over a year, including installation of new equipment.

By early 2007, the transition was largely complete, except for the instructional lab that housed the microfabrication course (ECE 557), which continued to operate in labs in the Electrical Engineering Building (now Brown Hall of Electrical Engineering) for several years. During the planning phases of the building, the instructional lab and associated course (ECE 557) were slated to be located on the first floor of the laboratory wing. However, during construction, it was decided to move the instruction lab to H Bay within the Scifres cleanroom. This location is still in use and slated to receive updated/upgraded equipment.

FACILITATING NEW USERS

Following the move-in phase, attention turned to the issues of facilitating new users and establishing an effective culture for the building. Tim Sands became the Birck director on November 1, 2006. Under his leadership, the number and strength of collaborative research interactions increased. Posters in the first floor atrium tracked the number of publications between research groups, which grew year by year. Several research groups effectively merged, including some partnerships involving faculty from different departments. Discovery Park was structured outside of the typical department/college silos, which lowered the barriers for interdisciplinary research. The Birck engineering staff increased in size and took on new roles related to facility operations, user training, and equipment expertise.

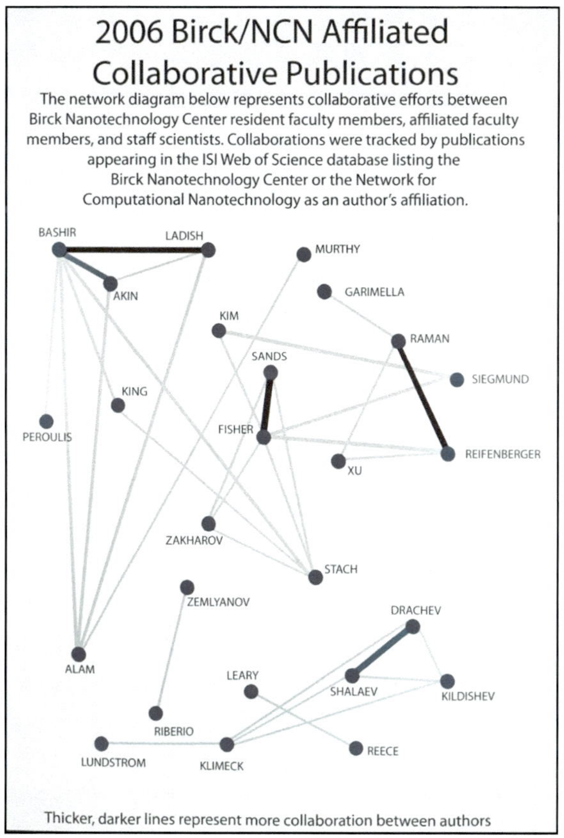

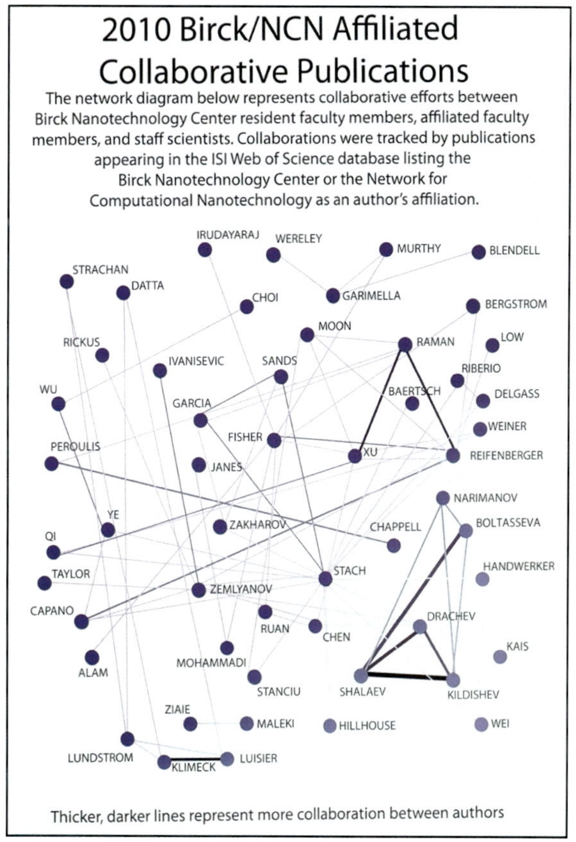

We were bringing all these people in from across campus, blending different cultures, protocols, and materials. I helped behind the scenes by working across disciplinary boundaries, collaborating with people, finding new ways of thinking and a common vocabulary.

David B. Janes
Co-Director of Semiconductor Degrees Program

Left: Posters showing collaborations among faculty groups in 2006 and just 4 years later in 2010.

Conference rooms.

Early on, it was determined that Birck needed to support projects in many technical areas and involving people from many disciplines. A one-size-fits-all set of policies wouldn't adequately support the user base. For example, clear/colorless liquids utilized in various labs ranged from water to solvents to acids and bases; each of these required a different set of handing protocols, personal protective equipment (PPE), and specific chemical hoods. A Policy and Procedures committee, including staff, faculty, and students, was formed. The P&P committee reviewed safety and PPE policies, with an aim of establishing guidelines for each set of labs that ensured safety and allowed effective monitoring of users. The committee also developed disciplinary policies and retraining requirements for users who violated operating and/or safety protocols. Overall, the P&P committee was successful in establishing a set of guidelines and practices to allow safe and effective use of the new multidisiciplinary facility.

NEW EQUIPMENT

New equipment was installed first, allowing staff members to gain confidence in shutting down and moving the equipment they currently used in other Purdue buildings to relocate it in Birck. Additionally, the support equipment—such as wet processing stations, all of which were new—would be installed prior to the equipment they served. This was also true for the characterization of equipment that would be required to be operational to verify the function of equipment after the move.

EQUIPMENT RELOCATED

Approximately half of the equipment to be moved into Birck was in operation in other Purdue buildings and supported active research projects. As a result, the time from when a piece of equipment was shut down until it returned to operation in Birck was critical. A complex plan to minimize this downtime was devised. Each engineer had only one piece of equipment moving at a time. The engineer acted as the construction manager for the equipment under their control and ran the project to relocate that piece of equipment. The system worked. Approximately 168 pieces of equipment were moved into Birck and installed.

Mark L. Voorhis was hired as a consultant and later building manager, holding that position today, because he had experience in moving research equipment. "John Weaver saw that my past experience would be a benefit to Birck because he was looking at moving roughly 170 tools to be installed in a very short amount of time," Voorhis recalled in 2025. "We brought tools from different buildings all over campus and under different conditions. We had to see what would fit into labs, what would fit into the cleanroom, and what needed to be modified."

> *I was the first person, first lab, to move in with the atomic force microscope. We moved from mechanical engineering. Nanotechnology was the next big-big thing, and the idea that we were putting together an interdisciplinary research facility—physicists, chemists, and folks from all kinds of disciplines together—was extremely appealing.*
>
> **Arvind Raman**
> John A. Edwardson Dean of the College of Engineering

> *Many comments were made about this $4 million move-in and installation project coming in on time and under budget. The most common was, "Neither of these things ever happen at Purdue, let alone both on the same project!" It became a source of professional pride for all those involved.*
>
> **John Weaver**
> Facility Manager, Birck Nanotechnology Center

FINANCIAL MILESTONES

Financial milestones were set, such as the number of research dollars procured in the first year, the first five years, and steadily after ten years. These milestones were used to justify the dollars that the university put into the Birck project. In all measurements, Birck Nanotechnology Center far exceeded the financial numbers and lessened the time for that goal to be reached.

Birck became an example of excellence for other nanotechnology facilities around the world. It was recognized with awards from international organizations, such as the prestigious Lab of the Year award and the *Cleanrooms* magazine Cleanroom of the Year award.

Collaboration in the Scifres cleanroom.

LEARNING TO WORK TOGETHER

OPERATION The operational methodology of Birck is unique. The concept of researchers securing membership in the facility to allow for shared equipment and spaces was a novel approach. Researchers were accustomed to buying and maintaining their own equipment. As an interdisciplinary facility with researchers from different departments, Birck is set up as a recharge center. Researchers pay for usage time on each piece of equipment using a "cost to cost causers" model for the most equitable method of determining fees. Birck was established with a zero-sum budget that breaks even at the end of each year, taking in just enough fees to cover costs.

SHARED SPACE AND EQUIPMENT Birck is based on the concept of a shared facility with shared equipment. No department or research group owns space in Birck. Equipment is open for use by all. A three-person team—staff engineer, faculty member, and senior student—oversees each piece of equipment.

TRAINING PROGRAMS AND COMMUNITY Extensive training programs were developed for trades, janitorial staff, and users of the facility. Deeper training is provided by the lab staff on each piece of equipment. Training on specific laboratory usage is provided, and access to the laboratory is dependent on successful completion of the specified training modules. Significant effort went into building a community among the staff, faculty, and users of the facility. Simple things like the purchase of a Keurig coffee maker for everyone's use, the celebration of holidays with carry-in meals, and other social events bring the atmosphere of a family into Birck.

GROWTH The number of users grew slowly at first, with a few pieces of equipment that required advanced building technology moving in, such as an atomic force microscopy (AFM). Other groups were held back by two factors: the lack of experience in a shared equipment setting; and the fact that researchers were charged for time spent on equipment.

However, it didn't take long before word spread from the initial faculty that the system worked, and the additional cost was well worth it. As faculty built the user fees into their grant proposals, use of the facility grew rapidly from about 50 percent occupancy to the point that areas of the facility were used for purposes other than for what they were originally designed. At that point, occupancy grew to about 120 percent of the design level.

With more researchers moving into the facility, the attraction of research dollars skyrocketed well above projected levels. The goal to attract world-class faculty was accomplished. Outstanding faculty members said they specifically came to Purdue University because of the Birck Nanotechnology Center.

Working at the FEI Titan transmission electron microscope (TEM) in Birck.

EXTENDING THE LAND GRANT LEGACY THROUGH NANOTECHNOLOGY

ALI SHAKOURI BECOMES DIRECTOR

Ali Shakouri was the director of Birck Nanotechnology Center from 2011 to 2021, following the headship of Tim Sands, who was director from 2006 to 2011. Sands left Purdue to become the president of Virginia Tech University, after serving as Purdue's provost and interim president. In 2025, Shakouri is a professor in Purdue's Elmore Family School of Electrical and Computer Engineering and associate dean of research for the College of Engineering.

As Birck's director for ten years, Shakouri has a deep knowledge of the center's past. In 2025, he explained why he was attracted to the position: "I was very excited about the job at Purdue because Birck is such a unique facility—not only the lab, but the people are outstanding and very collaborative. I worked closely with Tim Sands since 1999, and I knew quite a few faculty at Birck. I felt that there was potential for us to do more and have a deeper impact on science and engineering."

A VIBRANT ECOSYSTEM

In the early months under Professor Ali Shakouri's directorship, the overriding goal was finding how Birck could distinguish itself. He concentrated on how to build on the pillars of a world-class facility and strengths in nanoelectronics and photonics. It was apparent that the local ecosystem was a very important component. Professor Shakouri focused on how to create that ecosystem. Development included visiting nanotechnology centers at sister universities in the area, such as the University of Illinois, Notre Dame, Michigan, and Cornell, to understand their strengths and how best to leverage these unique facilities. After these visits, the focus turned to what was unique at Purdue and Birck.

There was an opportunity to leverage Purdue's strengths in professional colleges (Pharmacy, Ag and VetMed). For example, Purdue has top-notch faculty in drug development. This was the starting point for centers such as the Advanced Lyophilization Technology Hub (LyoHub) to bring engineering and pharmacy together in one facility. The goal of the LyoHub is to advance the science and technology of freeze-drying.

The other opportunity that came to the forefront was the College of Agriculture at Purdue. This led to collaboration with Karen Plaut (then associate dean of agriculture) to develop the roll-to-roll facility for low-cost printed smart films, which was highlighted during the visit of Secretary of Agriculture Tom Vilsack on November 17, 2014.

I was excited about joining the faculty at Purdue because of Birck, which is such a unique space, and also because the people are very collaborative.

Ali Shakouri
Birck Director, 2011–2021

Photo courtesy of WLFI TV18

Left, middle: Secretary of Agriculture Tom Vilsack visits Purdue.

ORIGIN OF THE LAND GRANT MISSION

In the early 1900s a dean of agriculture at Cornell realized that universities needed to be better connected to farmers and farm families. This idea led to the Smith-Lever Act of 1914 that created the cooperative extension service (today's Purdue Extension). The cooperative extension service connects each land grant university like Purdue to each county within its state. Research in the College of Engineering is less connected to local communities. This was a missing link that was bridged through the SMART Consortium and WHIN.

Dean Plaut thought it was important to showcase the possibilities within Birck that set this facility apart from others. Many universities had experimental agriculture farms and many had standard cleanrooms. This led to the formation of the SMART Consortium. Major agriculture and food companies who traditionally worked with the College of Agriculture now had a way to work more closely with Engineering and science faculty.

LAND GRANT UNIVERSITY MISSION REALIZED

Creating a powerful economic ecosystem around the center's strengths spurred a major project called the Wabash Heartland Innovation Network (WHIN), with Birck at the center of the initiative. Lilly Endowment conducted a study of the region that showed approximately 400,000 people living in the ten counties surrounding Purdue (approximately half of the population in Tippecanoe County). The study found that the two biggest economic drivers for the 9 surrounding counties were manufacturing and agriculture. Some of the best farmland in the country is in this region, but farmers are under a lot of economic pressure. The other strength of the area is the 200 to 250 manufacturers, companies such as Subaru, Caterpillar, Tate and Lyle, as well as many small manufacturers. This is unique. The third element is Purdue. To build a prosperous region, we had to bring the three strengths together.

This was 2017, and it was the first time I could see the problems of the rural counties. I could see why these counties were hurting. There were unemployment issues and drug issues, yet none of the manufacturers in our region had an R&D lab, and our (Purdue) faculty only worked with R&D labs. There were many farmers in the region, yet none were a high-tech experimental farm, so Purdue research didn't impact the surrounding community. Research funds given to Purdue were not helping the local community to utilize the intellectual strength of Purdue University.

Ali Shakouri | Birck Director, 2011–2021

Maxwell roll-to-roll manufacturing equipment.

When I first came to Purdue,
I never thought I would do anything
related to regional development,
but after seeing the challenges,
I felt it was important and fulfilled
the mission of Purdue as a land
grant university.

Ali Shakouri | Birck Director, 2011–2021

Birck directors at the tenth anniversary celebration: *from left,* Ali Shakouri (2011–2021), Richard Schwartz (2005–2006), James Cooper (2005–2006), Tim Sands (2006–2010).

Tenth anniversary celebration.

CREATING A LIVING LABORATORY

Birck had many low-cost Internet of Things (IoT) sensors, so it was suggested that those sensors be put to use in the ten-county region in both agricultural and manufacturing settings to create a living laboratory. This allowed research with real-time data and hopefully organizations using the sensors could improve their operations. This morphed into the WHIN project. Some of the knowledge generated from this work has been transferred to the agriculture extension program. This has also resulted in a National Science Foundation Future Manufacturing program.

BIRCK TENTH ANNIVERSARY CELEBRATION

The SMART Consortium was established during Ali Shakouri's directorship. Purdue's SMART—Scalable Manufacturing of Aware and Responsive Thin Films—Consortium is comprised of interdisciplinary researchers, collaborating with industry leaders, who are dedicated to answering one unifying question: How can IoT sensor technology become so affordable that it can be produced at mass volumes and deployed over large areas?

Low-cost, high-volume non-silicon and silicon sensors can transform numerous industries—from agriculture, manufacturing, sports, healthcare, and emergency response to food packaging, infrastructure, and sensor networking.

CREATING A CONFERENCE AS IF IT IS THE YEAR 2050

The year 2019 marked the 150th anniversary of Purdue, and the university welcomed the opportunity to celebrate many historic achievements. As the director of one of the most advanced research facilities in the world, Professor Ali Shakouri asked speakers from across all colleges to create talks, not about what the year 2050 would possibly be like, but to present as if it were 2050.

This led to innovative, artistic collaborations where the conference rooms were designed with bold futuristic ideas. Participants communicated via headsets, and food was served in pill form. Speakers gave futuristic talks on manufacturing, healthcare, housing, and electronics.

Conference 2050 exhibits.

As I gave tours, I pointed out the Scifres cleanroom, next to the Microsoft Quantum Lab, next to LyoHub. When asked what is the relationship to each other, I would say, "The relation is interdisciplinary technology solving problems in the society."

Ali Shakouri | Birck Director, 2011–2021

THE SHAKOURI LEGACY

During Ali Shakouri's tenure as the director of the Birck Nanotechnology Center, twenty new resident faculty from six colleges joined the center, and 13,000 square feet of lab space were renovated. Faculty research expenditures at Birck increased by more than 65 percent, and about $18 million was invested in shared new tools. Interdisciplinary research continued to be strong.

The Birck faculty have worked in a collaborative way for the good of the center, even relinquishing their own lab space when new technologies and opportunities were presented, helping to acquire world-class researchers in materials and characterization. The flexibility of the building itself made transitions possible.

The Birck Center contributed greatly to the College of Engineering's preeminent teams, which identified nine key areas that were determined to be a great strength at Purdue. Four of the nine teams were greatly impacted by the exisitence of Birck. These teams led to the hiring of 100 new faculty members.

***Right*: Conference 2050 participant trying out communication headset.**

Purdue 2050
Petronio
Bendito
Purdue University

ACCELERATING INNOVATION TO IMPACT

A ONCE-IN-A-GENERATION OPPORTUNITY

BY DR. ZHIHONG CHEN

I stepped into the role of interim director of the Birck Nanotechnology Center in 2022—the same year the CHIPS and Science Act was enacted—during a pivotal moment for the nation's semiconductor strategy. In 2023, I was honored to be appointed director with a clear objective: to ensure that Purdue, and Birck in particular, would be well-prepared to contribute meaningfully to this once-in-a-generation opportunity and respond to emerging national priorities.

As the largest and most advanced academic cleanroom and characterization facility in the country, Birck was already a critical asset. Recognizing its strategic role, Purdue's Board of Trustees approved a $49 million renovation plan, set to begin in 2024. These upgrades are designed to support semiconductor, advanced packaging, and quantum research, enabling the continued development of a robust innovation ecosystem.

In conjunction with infrastructure support from the Silicon Crossroads Microelectronics Commons (SCMC) Hub and the Lilly Endowment, as well as partnership with the Semiconductor Manufacturing and Advanced Research with Twins USA (SMART USA) Institute, more than $75 million has been invested in the facilities upgrades and new tools for processing and metrology. Birck is well-positioned to advance the CHIPS Act's goals—particularly in accelerating the lab-to-fab transition and supporting cost-effective prototyping of novel materials, devices, and processes in domestic facilities.

Over the past three years, Birck has gained growing recognition for its technical capabilities and role as a platform for collaboration. The center has welcomed numerous visitors—including industry leaders, federal and state officials, and international delegations—who have engaged with our efforts in research, development, and workforce training.

Birck also plays an integral role in Purdue's Semiconductor Degrees Program, which supports the national effort to train over one million additional skilled workers by 2030 and triple the number of STEM graduates. The center offers hands-on learning opportunities in both cleanroom and characterization laboratories. In Spring 2023, we launched the Semiconductors@Birck Vertically Integrated Projects (VIP) course to provide undergraduate students with access to semiconductor fabrication and introduce key concepts such as statistical process control and advanced packaging. Programs like the STARS summer camp further strengthen experiential learning and help cultivate student interest in microelectronics careers.

The current renovation marks more than an update to Birck's 20-year-old infrastructure. It represents a significant expansion of our capabilities, including the addition of 8-inch wafer-compatible toolsets for patterning, deposition, etching, annealing, and packaging, complemented by comprehensive characterization cores. These investments are enabling Birck to better serve as a research and education facility that supports both near-term innovation and long-term national goals.

Celebrating the first phase of $100M enhancements—cutting a ceremonial ribbon in front of the Scifres Nanofabrication Laboratory. *Left to right*: Director Zhihong Chen, then-NSF Director Sethuraman Panchanathan, U.S. Senator Todd Young, President Mung Chiang, then-EVPR Karen Plaut.

I'm very proud of the people that I work with and the folks that I've come in contact with. This building was designed very well. I recognized that when I was first hired in 2004, and I wanted to make sure I didn't do anything to violate the original design, because I thought the original design was excellent.

Mark L. Voorhis
Birck Nanotechnology Center
Building Manager

BIRCK MODERNIZATION PROJECT

Even with modifications to the building to accommodate more laboratory space, the efforts in the facility became limited by the size of the building. After twenty years of operation, it was clear that something dramatic needed to be done to further growth in research.

Zhihong Chen, the Reilly Professor of Electrical and Computer Engineering at Purdue, became the fourth Mary Jo and Robert L. Kirk Director of Birck Nanotechnology Center in 2023, following one year of service as interim director. On April 14, 2023, Purdue's Board of Trustees approved $49 million to Birck for a facility upgrade, known as the Birck Modernization Project. Thirty million was designated for the building's systems, such as upgrades to air handling and ultrapure water infrastructure, as well as a major redesign to expand the amount of filtered space in the cleanroom.

Nineteen million dollars were earmarked for the upgrade of equipment, including a full suite of tools that accommodate 8-inch diameter wafers, enhancing the industry relevance of Birck's research and facilitating greater collaboration. Birck is at the forefront of the transition to 8-inch tools, setting it apart from most other university facilities. These upgrades not only enhance research capabilities but also strengthen Birck's role as a collaborative platform for lab-to-fab transition, allowing industry partners to prototype and scale innovations in a university setting with industrially relevant infrastructure.

There was a need for more cleanroom space, fewer return air chases, and basically that's what we've been doing over the last year and a half for upgrades. We replaced obsolete air compressors. We upgraded the ultrapure water system to two times the capacity—from 100 gallons a minute to 200 gallons a minute—to help take care of our future demands. The design of the facility was made in such a way that there are galleys down the middle between two rows of research labs. The galleys are where the pumps and things of this nature that can contaminate the research tools are located. The building was very well designed from a power standpoint. We ran out of space before we ran out of power, which is a good thing. Normally, it's the other way around.

Mark L. Voorhis | Birck Nanotechnology Center Building Manager

Air handling system parts arrive for the 2024 renovation.

NEW ADVANCED PACKAGING AND QUANTUM MATERIALS CHARACTERIZATION FACILITIES

The upgrades also created a dedicated cleanroom and lab space for advanced packaging research—innovations in design, materials, and processing that are needed to integrate chiplets (small, modular chips designed to be integrated into a larger system) with different functionalities—and shared quantum transport and characterization facilities, which are important to future electronic and quantum information science applications.

QUANTUM RESEARCH AND MICROSOFT QUANTUM AT BIRCK

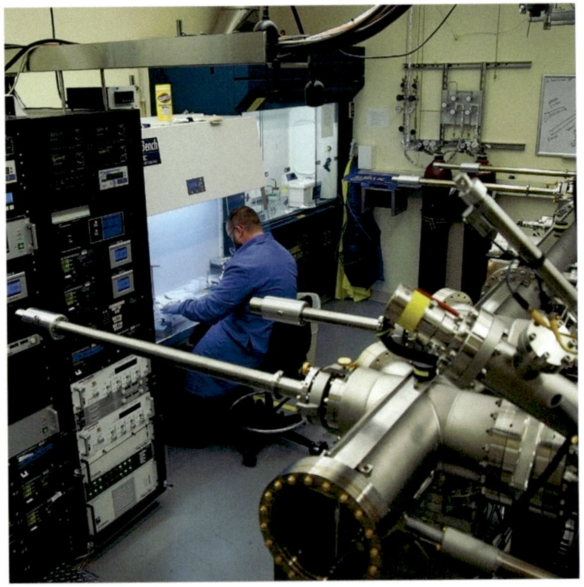

MBE Lab.

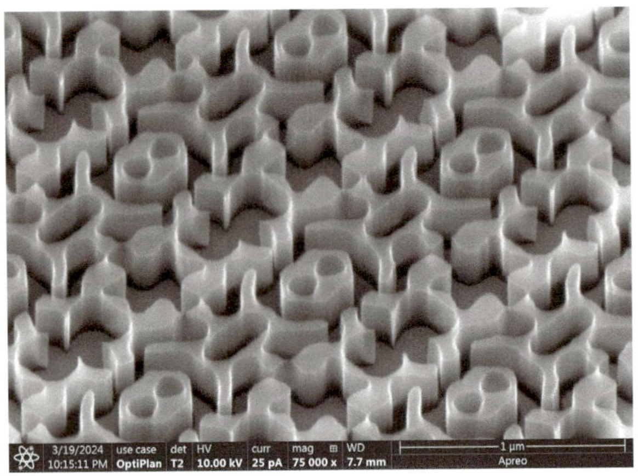

Electron microscope image of Birck-designed nanophotonic dielectric resonators etched in Si on sapphire, fabricated in the Scifres cleanroom and studied in the nanophotonics labs at Birck.

The last several years have witnessed a rapid expansion of activities in quantum science and engineering at Birck. The physical infrastructure and expertise of the research engineering staff at Birck have made it an obvious location for activities in quantum science research and development of new quantum technologies. Many faculty members spanning the College of Science and the College of Engineering are engaged in pioneering work in quantum science at Birck, including strong efforts in quantum photonics and communications, quantum sensing, novel quantum materials growth and characterization, and quantum device fabrication and testing.

The Purdue Quantum Science and Engineering Institute (PQSEI) is a central organizing body for quantum research at Purdue with over 65 faculty members participating. PQSEI works to establish collaborations among Purdue faculty, and more broadly with industrial and governmental partners, to transform innovations in basic quantum science into useful technologies. PQSEI leadership is stationed in the Birck Center, facilitating coordinated efforts for federal funding opportunities.

Recent successes include the DOE Quantum Science Center (QSC, 2020–2025) led by Oak Ridge National Laboratory, one of five such centers funded nationally by the Department of Energy. Purdue University serves as the primary academic partner for Oak Ridge with twelve Purdue faculty members serving as co-investigators. Much of the Purdue QSC activities take advantage of the resources available at Birck. Another example is the Center for Quantum Technologies (CQT), a Nation Science Foundation–funded center that brings academic researchers together with industrial partners interested in the development of quantum technologies to work on projects of mutual interest.

One of the most visible of several examples of Purdue's leadership in quantum computing is found in Microsoft Quantum Lab West Lafayette. Microsoft Quantum Lab West Lafayette is an integral part of Microsoft's global effort to build a full-stack utility scale quantum computer exploiting the principles of topological quantum computing. At Microsoft Quantum Lab West Lafayette, Microsoft engineers and scientists work to develop quantum materials and devices that form the backbone of the quantum plane of a full-stack quantum computer.

Microsoft Quantum leverages the infrastructure and expertise available at the Birck Center to achieve its goals. This fully fledged Microsoft group is an outgrowth of a long-standing collaboration with Professor Michael Manfra to develop the quantum devices necessary to build a large-scale quantum computer. This thriving partnership between Purdue and Microsoft is approaching its tenth year of operation. The available infrastructure and expertise existing in the Birck Center was a foundational aspect of bringing Microsoft to the Purdue campus for such a large-scale effort. We have added state-of-the-art molecular beam epitaxy growth capabilities and advanced lithography tools to the Birck Center as a consequence of this project. Microsoft leverages advanced material and device characterization tools available at Birck as well, making for a productive research environment that exposes Purdue postdoctoral fellows and graduate students to the front lines of technology development.

> *One of the principal attractions for me to move to Purdue fifteen years ago was the Birck Nanotechnology Center. I was moving from an industrial research laboratory, Bell Labs in Murray Hill New Jersey, and I wanted to find an academic home that could give me an opportunity to recreate in some way what I had at Bell Labs. The Birck Center was clearly a place where I felt I could realize my research goals, and the Purdue administration understood and appreciated what I hoped to achieve. The Birck Center provided me a great opportunity and has played an enormous role in what has been established today in the quantum space.*
>
> **Michael Manfra** | Bill and Dee O'Brien Distinguished Professor of Physics and Astronomy, Director of PQSEI, and Director of Microsoft Quantum Lab West Lafayette

BNC's unique multidisciplinary environment as envisioned by Birck's founding fathers enables truly novel research at the intersection of different disciplines and has been transformative for our quantum photonics research merging optics, nanoscience, computations, and quantum physics.

Vladimir Shalaev | Robert and Anne Burnett Distinguished Professor of Electrical and Computer Engineering

Alexandra Boltasseva | Ron and Dotty Garvin Tonjes Distinguished Professor of Electrical and Computer Engineering

NATO Deputy Secretary General Mircea Geoana experiencing Birck cleanroom virtual fabrication with the Apple Vision Pro.

INDUSTRY PARTNERSHIPS: ACCELERATING THE LAB-TO-FAB TRANSITION

As Birck celebrates its twentieth anniversary in 2025, several innovative initiatives are underway, guided by the center's updated mission: Educate, Innovate, Collaborate. The renovation and addition of 8-inch wafer-scale fabrication and state-of-the-art characterization capabilities—pushing the boundaries of spatial and energy resolution—are enabling groundbreaking discoveries and innovation.

The new wafer-scale tools and the unprecedented implementation of statistical process control (SPC) in a university cleanroom setting have made Birck an attractive partner for industry collaborators seeking advances in materials, devices, and processes. Major partnerships have been built with imec (Interuniversity Microelectronics Centre), the world's largest microelectronics R&D center, which works with over 600 industry partners. This collaboration combines imec's deep lab-to-fab expertise with Birck's agile innovation platform, allowing industry-aligned prototyping and accelerating the path from concept to market.

The partnerships with fabs and foundries ensure that Birck's research is industry-relevant, supports time-to-market goals, and strengthens semiconductor workforce development through curricula and training materials validated by industry needs. Supply chain companies are also drawn to Birck for materials and process innovation because the facilities are uniquely flexible, allowing the testing of new materials and processes while maintaining rigorous statistical performance and quality metrics.

DIGITALIZATION OF THE CENTER

Marking its twentieth year, Birck has launched efforts to integrate artificial intelligence and machine learning (AI/ML) and digital twin technologies across its physical facilities, education programs, and workforce training initiatives. Digital twin technology involves creating a virtual replica of a physical object, process, or system that is continuously updated with real-time data, enabling simulation, analysis, and optimization. These digital twins enhance understanding of complex systems, offer predictive insights, and support proactive decision-making to improve performance and reliability. In parallel, Birck is developing an AI-powered chatbot for equipment training, designed to provide personalized, user-paced learning experiences that improve safety, accelerate onboarding, and enhance user confidence in cleanroom operations.

> *We are incorporating digital twins and AI-driven—or even autonomous—R&D into both our infrastructure and daily operations. A key initiative is the development of a virtual Birck, which will allows students to learn and practice semiconductor fabrication and characterization in a virtual environment before entering the cleanroom. This immersive approach democratizes access, enhances safety, boosts confidence, and accelerates learning. By enabling simulation, process iteration, and data analysis in the virtual space, we will significantly shorten R&D cycles and lower overall research costs.*
>
> **Zhihong Chen** | Mary Jo and Robert L. Kirk Director of the Birck Nanotechnology Center

My vision is to make Birck the nation's leading virtual twin-enhanced nanofabrication facility—where the physical and virtual worlds converge to shorten R&D cycles, reduce cost, and accelerate innovation. What makes Birck uniquely positioned is not only our advanced capabilities, but the unified support from the university and the strength of our industry partnerships.

Zhihong Chen
Mary Jo and Robert L. Kirk Director of the Birck Nanotechnology Center

With Purdue as one of the founding members of the SMART USA Institute and in partnership with Dassault Systèmes and other companies, Birck, together with Purdue's nanoHUB and the Institute for Physical AI (IPAI), is building an infrastructure for automated data collection, streaming, and management to support AI/ML modeling integrated with physics- and chemistry-based simulations. These models are designed to evolve into responsive, surrogate-enabled digital twins.

The vision is to create a digital twin–enhanced facility that democratizes access to semiconductor fabrication through immersive virtual reality/augmented reality (VR/AR)–based training, supports sustainable cleanroom operations through real-time monitoring and optimization, and enables accurate simulation of unit processes and short-loop flows. This integrated approach is designed to accelerate process development, shorten design-of-experiment cycles, and bring greater agility to lab-to-fab transitions.

Over the years, I have grown more and more to appreciate Birck for the design of the facility. Not just in terms of the research capabilities but also for the positive culture environment for me during the nascent days of the center. I've kept track of others who have joined and the work that has taken place. It continues to be impressive. I don't know if there's another campus across the country that has been such a successful force for the boom of a semiconductor refocus. I loved that place. It's great that to this day it draws others to not just be there for the incredible academics, but to be there because of the capabilities at Birck as a key part of the institution.

Aaron Franklin
PhD from Purdue University in 2008; currently the Addy Professor of Electrical and Computer Engineering and Associate Dean for Faculty Affairs at Duke University

MIDWEST UNIVERSITY NETWORK FOR NANOCENTERS

As part of SCMC's R1 Nucleus (R1N), Birck plays a central role in shaping a future Midwest university nanocenter network. Together with peer institutions University of Illinois Urbana-Champaign, Notre Dame, and University of Michigan, the R1N strengthens collaboration across leading academic cleanrooms, fosters shared infrastructure and standards, and creates a more cohesive regional platform for lab-to-fab research, workforce development, and semiconductor ecosystem support. Reciprocal agreements among the four university facilities allow users to access all sites without redundant training, while benefiting from the complementary capabilities each facility offers.

MARK LUNDSTROM, ONLY CHIEF SEMICONDUCTOR OFFICER IN US

I have this unusual title. I'm the chief semiconductor officer for Purdue, and I'm quite certain I'm the only chief semiconductor officer in the US. I think it's a reflection of the commitment that Purdue has as one of the largest STEM universities in the country. We feel it's our responsibility to help the nation with the most significant challenge it faces right now. So, I'm working with an incredibly talented and hardworking group of faculty on a lot of different initiatives having to do with workforce and economic development, while also working with a team in the state to bring more of this industry to Indiana.

Mark Lundstrom | Chief Semiconductor Officer, Don and Carol Scifres Distinguished Professor of Electrical and Computer Engineering, Purdue University

BIRCK AND THE CHIPS ACT

The semiconductor work done by Purdue and Birck Nanotechnology Center is gaining increasing prominence because of the bipartisan federal CHIPS and Science Act of 2022. CHIPS aims to boost domestic research and development of semiconductors and microelectronics. The CHIPS Act provides funding and incentives for semiconductor research, development, manufacturing, and workforce development in the United States, and more importantly, it fosters strong partnerships between academia and industry.

WHAT IS A SEMICONDUCTOR?

> *Most people know that metal conducts electricity, and an insulator does not conduct electricity—like glass. A semiconductor is somewhere in between. So, what's the big deal? The big deal is that by introducing small amounts of other atoms in a semiconductor crystal, you can dramatically change its properties to make it more metallic, or more semiconducting, and that's how we make electronic devices like the transistor, which is the basic building block of all electronics. It makes everything possible that we think about today. The smartphones we carry around are chock-full of chips. There are a few thousand chips in every automobile. It's hard to imagine life without semiconductors.*
>
> **Mark Lundstrom** | Chief Semiconductor Officer, Don and Carol Scifres Distinguished Professor of Electrical and Computer Engineering, Purdue University

Semiconductors are a vital component of devices that Americans use every day—including smartphones, cars, and laptops—as well as the military equipment that protects our country. Yet the United States, which spearheaded development of semiconductors, now makes only 10 percent of them, and not the most advanced ones. About three-quarters of all semiconductors are produced in East Asia. Anchored by the Birck Nanotechnology Center, Purdue is well-positioned to advance U.S. leadership in semiconductors through cutting-edge research, close collaboration with co-located industry partners, and hands-on training for students.

Top: Soon after the signing of the CHIPS Act in 2022, Birck welcomed distinguished visitors that demonstrated the center's national prominence, including then-Secretary of State Antony Blinken and Commerce Secretary Gina Raimondo.

Left: U.S. Senator Todd Young and NSF Director Sethuraman Panchanathan taking selfies with Purdue President Mung Chiang.

ONE OF THE FIRSTSEMICONDUCTOR DEGREE PROGRAMS

The Birck Nanotechnology Center plays a vital role in developing the country's high-tech workforce of tomorrow. Purdue is the home of the nation's first comprehensive semiconductor degrees program established in 2022. A full suite of offerings includes graduate and undergraduate degrees, competitive internships, flexible online courses, and a virtual lab. Much of the hands-on training takes place at Birck, putting students at the center of emerging Silicon Heartland technology efforts.

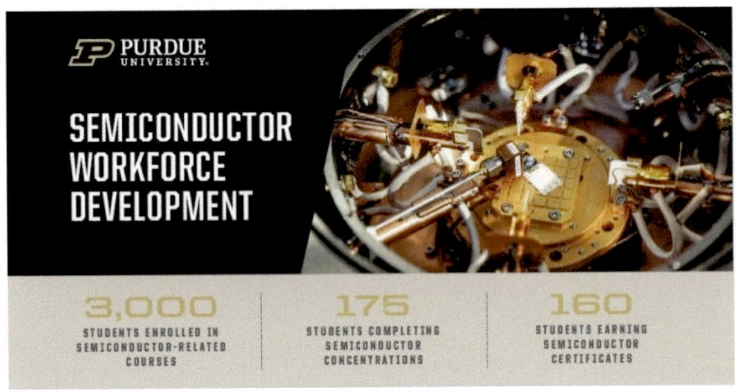

STARS AND VIP PROGRAMS @ BIRCK

Purdue's company-sponsored STARS program introduces students to semiconductors through intensive, industry-relevant projects in Birck's cleanroom and labs. The new STARS 2.0, supported by the Lilly Endowment and other industries, will expand capacity and add academic-year components. Students then advance to Semiconductors@Birck, a Vertically Integrated Projects (VIP) program launched in Spring 2023 and recognized by industry fabs as a premier training pathway for future process, equipment, and integration engineers. Through multisemester projects, they gain nano-fabrication expertise, teamwork, and project-management skills while contributing to Birck initiatives such as statistical process control (SPC) and advanced packaging. Most recently, the Virtual Twins@Birck VIP launched an effort to build digital twins of the Scifres cleanroom and its equipment, directly supporting SMART USA Institute's mission to democratize semiconductor education and training.

> *VIP students gain hands-on experience along with skills for professional development. It's not only a course to teach knowledge and skills, but also a course that gives students the platform to perform teamwork. This type of professional development is critical for their future careers. The inaugural ten students in the first semester loved the VIP course and word spread quickly. We were able to recruit thirty students in the second semester, and eventually we had fifty students maximum because of the capacity of our cleanroom. Many employers appreciate the strength of our students and the quality of our educational programs. This is a success story for Birck's expanded role in education. We are utilizing our facility and providing a unique opportunity for undergraduate students.*
>
> **Zhihong Chen** | Mary Jo and Robert L. Kirk Director of the Birck Nanotechnology Center

Semiconductors@Birck program participants.

> *In the semiconductor field, workforce development is a huge issue right now in the sense that the industry is telling us they need tens of thousands of new employees over the next few years, maybe hundreds of thousands. We have a long tradition of having microfabrication lab courses and opportunities for students to work on projects. We had done it informally for a long time with students who would come by and say, "Hey, I want to work on a project." The Birck faculty integrates undergrads in their work. I think it's great.*
>
> **David B. Janes** | Co-Director of Semiconductor Degrees Program

DEFENDING THE U.S. AT SCALE

Scalable Asymmetric Lifecycle Engagement (SCALE) is an immersive educational program that combines government and industry internships with aligned research and mentoring to deepen understanding and relationship building. A public-private-academic partnership (PPAP) consortium-based approach provides national reach to 33 U.S. universities full partners and three affiliate partners in targeted semiconductor technical areas for scaling and replication of the program, while allowing a regional focus. SCALE combines inputs from all PPAP sectors; a set of national curriculum standards aligned with U.S. Department of Defense requirements in specialty areas that go beyond current ABET practice; a cohesive messaging and recruiting approach tailored for students from K–12 through PhD; an industry-supported Center for Secure Microelectronics Ecosystem (CSME) to engage students in transitioning research to missions; iterative adaptation of curricula and standards; and systematic programmatic evaluation to ensure increased effectiveness in meeting government and industry workforce goals. Both SCALE and CSME have strong ties to Birck, contributing to its ecosystem of education, innovation, and collaboration.

The Birck facility has a couple of great resources. One is the facility itself. It's a unique space on campus. We were looking recently to set up an undergrad microfabrication lab, and, other than the instructional bay in Birck, it is hard to find any other place on campus to do that. Birck also has a really good staff. These resources help to build an increased emphasis on the educational role.

David B. Janes | Co-Director of Semiconductor Degrees Program

8

LIFE AT BIRCK

NANOTECHNOLOGY STUDENT ADVISORY COUNCIL (NSAC)

The Nanotechnology Student Advisory Council (NSAC) is a student-run organization for graduate students interested in nanotechnology. The organization promotes collaboration, enhances research awareness, and facilitates communication among faculty, staff, and students. To meet these goals, they plan regular social events, host lab open houses, and collaborate with the staff to organize the Equipment & Process Expert (Super-User)

NSAC picnic.

Program. Their student volunteers serve on the Policies and Procedures Committee for Birck. The NSAC team also provides a tremendous boost for Birck community outreach, from conceiving and developing nanoscience activities to representing Birck at the outreach events.

OUTREACH

K-12 PROGRAM Birck holds a variety of engaging events and facility tours to inspire tomorrow's nanotech engineers and nanoscience researchers. A high priority of Birck's mission is to develop students' excitement about technology when they are young.

MANUFACTURING WEEK Birck brought its NanoDays outreach to the City of Lafayette's Manufacturing Week starting in October 2023, introducing students in grades K-5 to nanotechnology. Birck graduate researchers demonstrated the nanoscience of surface area, ferrofluids, and structural color. Meanwhile, middle schoolers received hands-on experience working with Birck research groups, including a chance to operate a scanning electron microscope and build a box camera. High schoolers toured Birck for a close-up look at the Scifres Nanofabrication Laboratory's cleanroom.

NANODAYS Birck's main yearly outreach event (pre-COVID) was hosting a two-and-a-half-day open house in our facility for over 1,000 children in grades K-8. The annual event was powered by a team of over 100 volunteers from among our Birck staff and residents, unique research demos from Birck faculty and their graduate students, and more than 20 NanoDays activities kits developed by the National Informal STEM Education (NISE) Network.

Middle school students at Manufacturing Week.

NanoDays K-4 outreach event.

GRANDPARENTS UNIVERSITY

Birck participates every year in Purdue's Grandparents University, hosting the All the Small Things nanotechnology major for children and their Purdue alumni grandparents. Two days of nanoscience activities were conducted in and around the Scifres Nanofabrication Laboratory cleanroom.

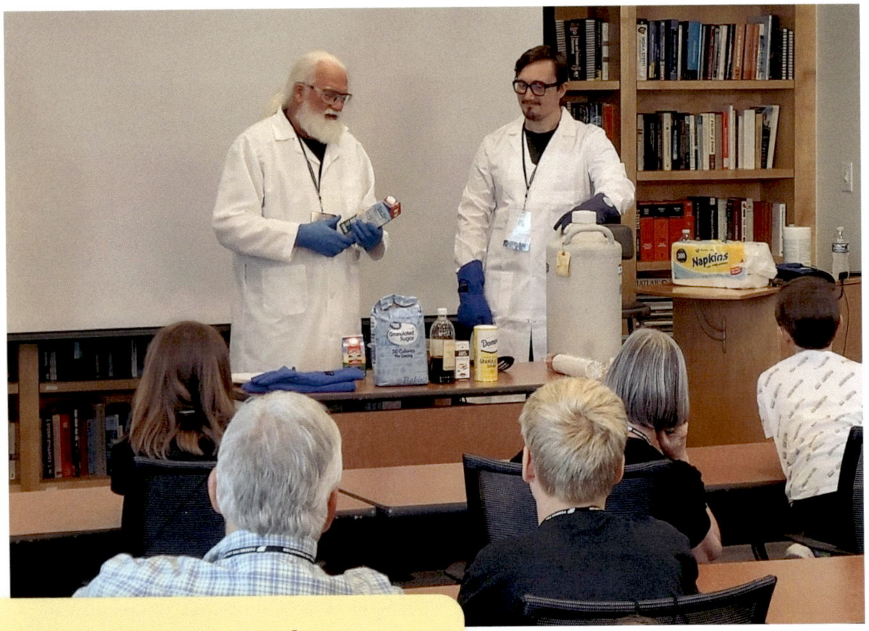

Birck staff making nano ice cream.

Nano Ice Cream Recipe

One quart half-and-half
One pint heavy whipping cream
One cup sugar
One tablespoon vanilla
Liquid nitrogen as needed,
about one gallon

Grandparents inside the cleanroom.

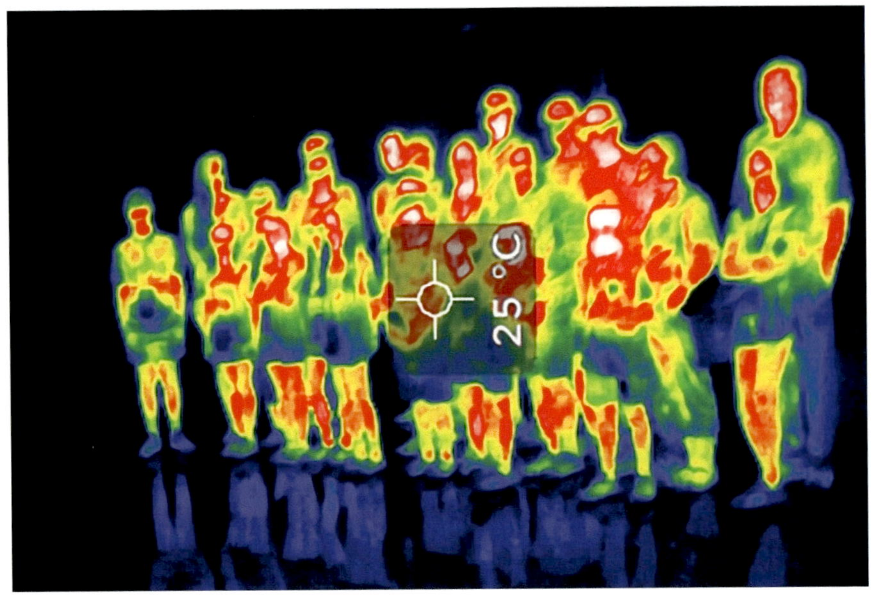

Grandparents University participants thermal image.

Birck holiday party.

Birck Thanksgiving party.

There is no doubt that the Birck Nanotechnology Center stands as a preeminent research facility, home to some of the world's leading faculty and student researchers. Yet beyond its academic excellence, Birck fosters a unique sense of community—most notably within its dedicated staff. The camaraderie among this team is unparalleled; they support one another through personal challenges and celebrate each other's milestones. A facility like Birck can only achieve greatness through the commitment and daily efforts of the people who bring its mission to life. That pride and dedication are evident across the entire staff—from engineers and research scientists to the building support team, custodial staff, administrative personnel, and business operations.

Tracy Hudson
Managing Director, Birck Nanotechnology Center

Wall mural on Birck's second floor atrium featuring submissions from the Birck community for a research image contest held in December 2019.

As we celebrate Birck's 20th anniversary, we look toward the future with renewed energy and purpose. With state-of-the-art facilities supporting advanced nanofabrication and characterization, Birck is uniquely positioned to drive cutting-edge research in semiconductors, quantum computing, information science, and beyond. Guided by our new mission of education, innovation, and collaboration, we are committed to shaping the next generation of technological breakthroughs. Meanwhile, we are leading the development of AI and virtual twin technologies with our strategic partners to transform how we educate, innovate, and conduct research—laying the groundwork for a more connected, intelligent, and impactful future.

Zhihong Chen | Mary Jo and Robert L. Kirk Director of the Birck Nanotechnology Center

DIRECTORS

- James Cooper, De Facto Leader, 1996–2001
- James Cooper and Richard Schwartz, Founding Co-Directors, 2001–2006
- Tim Sands, 2006–2010
- James Cooper, Interim Director, 2010–2011
- Ali Shakouri, 2011–2021
- Zhihong Chen, Interim Director, 2022; Director, 2023–present

SOURCES

Aaron Franklin interview with Tracy Hudson, 6/5/2025

Ali Shakouri interview with Zhihong Chen and Tracy Hudson, 5/21/2025

Arvind Raman interview with Zhihong Chen and Tracy Hudson, 6/5/2025

David B. Janes interview with Tracy Hudson, 5/12/2025, 5/19/2025

Family of Richard Schwartz interview with Tracy Hudson, 6/12/2025

James Cooper interview with Zhihong Chen and Tracy Hudson, 5/9/2025

John Weaver discussions with Tracy Hudson, 3/20/2025, 3/25/2025, 4/3/2025 (John provided a fair amount of written content)

Mark Voorhis interview with Tracy Hudson, 6/6/2025

Ron Reifenberger interview with Tracy Hudson, 5/12/2025

"Birck Nanotechnology Center," Purdue University Office of Research, accessed August 15, 2025, https://www.purdue.edu/research/oevprp/institutes-and-centers/facilities/birck.php

Emile Venere, "Purdue Launches Discovery Park, Nanotechnology Center," *Purdue News*, November 2001, https://www.purdue.edu/uns/html4ever/010907.Discovery.allinone.html

Emile Venere, "Purdue's Birck Nanotech Center to Be Dedicated Oct. 8," *Purdue News*, September 30, 2005, https://www.purdue.edu/uns/html3month/2005/050930.Adams.birck.html

"Birck Nanotechnology Center to Receive Significant Upgrade," Elmore Family School of Electrical and Computer Engineering, accessed August 18, 2025, https://engineering.purdue.edu/ECE/Wavelinks/2023/birck-nanotechnology-center-to-receive-significant-upgrade

"Prof. Ali Shakouri Appointed Purdue Engineering Associate Dean for Research and Innovation," Elmore Family School of Electrical and Computer Engineering, January 10, 2024, https://engineering.purdue.edu/ECE/News/2024/prof-ali-shakouri-appointed-purdue-engineering-associate-dean-for-research-and-innovation

"The SMART Consortium," Scalable Manufacturing of Aware and Responsive Thin Films, accessed August 18, 2025, https://engineering.purdue.edu/SMART

Birck website: https://birck.research.purdue.edu/

ACKNOWLEDGMENTS

As the idea to capture the history of the Birck Nanotechnology Center took shape, it quickly became evident how much personal and professional pride so many individuals feel for this remarkable facility. I have often thought of Birck as a family, and through this project, it became clear that many others share that same sentiment.

The building itself is a masterpiece—an architectural and engineering benchmark for university research facilities. Yet, it is the people who make it truly special.

More than twenty-five individuals contributed to this book, and I extend my sincere thanks to each of them for their thoughtful input and enthusiasm throughout this process.

I would especially like to acknowledge James Cooper, David B. Janes, John Weaver, Ron Reifenberger, Ali Shakouri, Mark Voorhis, Neil Dilley, Kelli Dawson, and Tracy Hudson. Their sustained engagement and substantial editorial contributions were instrumental to the success of this project. Their efforts went well beyond what was expected, and I am deeply grateful for their dedication and support.

The majority of the content of chapters 1, 2, 3, and 5 was compiled by James Cooper, John Weaver, and David B. Janes.

This book is a reflection of the enduring spirit of innovation, collaboration, and community that has defined the Birck Nanotechnology Center since its inception—and will continue to shape its legacy for generations to come.

Zhihong Chen | Mary Jo and Robert L. Kirk Director of the Birck Nanotechnology Center

INDEX

Page numbers in italics indicate photos

ABOUT THE AUTHORS

Zhihong Chen is the Reilly Professor of Electrical and Computer Engineering at Purdue University and the Mary Jo and Robert L. Kirk Director of the Birck Nanotechnology Center. Prior to joining Purdue in 2010, she was a research staff member and manager at the IBM Thomas J. Watson Research Center. Her research focuses on novel materials and devices for beyond-CMOS technologies. She served as associate editor of ***IEEE Electron Device Letters*** and held leadership roles in major conferences. An IEEE Fellow, she has published ninety-seven journal articles and sixty conference papers, holds thirteen U.S. patents, and has more than 19,900 citations to her name.

Angie Lipp Klink has authored fourteen books, several about the history and people of Purdue University. Her most recent book, ***Purdue's Female Founders: The Untold History of Trailblazing Women Faculty***, was published by Purdue University Press. She is an independent scholar, essayist, scriptwriter, and advertising copywriter. Klink has written for the American Writers Museum, ***Ms.*** magazine, ***Traces of Indiana and Midwestern History*** magazine, and documentaries narrated by actor Peter Coyote. She holds sixty-two American Advertising Federation ADDY Awards, including a Lifetime Achievement honor, and an Honorable Mention in the Erma Bombeck Writing Competition. Klink has written about author, poet, and artist Evaleen Stein, and procured a 2024 Indiana Historical Bureau marker in Lafayette, Indiana, honoring Stein. Klink is a recipient of the Excellence in Historic Preservation Medal from the National Society of the Daughters of the American Revolution, the organization's most prestigious award honoring dedication to the research and writing of books. Klink holds a BA in communication from Purdue University. Learn more at www.angieklink.com.